Risk Engineering

Series editor

Dirk Proske, Vienna, Austria

The Springer Book Series *Risk Engineering* can be considered as a starting point, looking from different views at Risks in Science, Engineering and Society. The book series publishes intense and detailed discussions of the various types of risks, causalities and risk assessment procedures.

Although the book series is rooted in engineering, it goes beyond the thematic limitation, since decisions related to risks are never based on technical information alone. Therefore issues of "perceived safety and security" or "risk judgment" are compulsory when discussing technical risks, natural hazards, (environmental) health and social risks. One may argue that social risks are not related to technical risks, however it is well known that social risks are the highest risks for humans and are therefore immanent in all risk trade-offs. The book series tries to cover the discussion of all aspects of risks, hereby crossing the borders of scientific areas.

More information about this series at http://www.springer.com/series/11582

Rita Yi Man Li

Construction Safety and Waste Management

An Economic Analysis

Springer

Rita Yi Man Li
Department of Economics and Finance /
Sustainable Real Estate Research Center
Hong Kong Shue Yan University
Hong Kong
China

ISSN 2195-433X ISSN 2195-4348 (electronic)
Risk Engineering
ISBN 978-3-319-12429-2 ISBN 978-3-319-12430-8 (eBook)
DOI 10.1007/978-3-319-12430-8

Library of Congress Control Number: 2014952913

Springer Cham Heidelberg New York Dordrecht London

Printed on acid-free paper

Springer International Publishing AG Switzerland is part of Springer Science+Business Media (www.springer.com)

Preface

Construction waste and safety are often perceived as two different issues by construction practitioners in Australia as they are managed by two different teams or persons in large construction sites. Nevertheless, the two issues are often put in the same tool box talks everyday with probably ten times more on safety as compared to waste. In Hong Kong, construction waste and safety are managed by the same person in a small or medium-sized construction site. Or they may be handled by two different teams if it is a mega project.

No matter how safety and waste are managed on sites, it is undeniable that there are opportunity costs in construction waste and safety management. As resources for managing safety and waste are limited, spending more on one item will leave less for others. In view of the close relationship between these two, this book perceives the issues from an economic perspective.

Chapter 1 studies the motivation behind sustainable construction waste management in Australia with regard to positive and negative incentives, goal setting and hierarchy of needs theory. Chapter 2 reviews comments from scholars with regards to construction and municipal waste in Hong Kong. Chapter 3 analyzes the static digital information about the causes of construction accidents. Chapter 4 evaluates big data problems with respect to construction safety index. Chapter 5 perceives the use of mobile apps in construction safety knowledge sharing from an institutional economic perspective. Chapter 6 reveals the supply and demand of safety regulations in China, Nigeria, Kuwait, and Ghana. Chapter 7 provides an analysis of construction safety motivations from the psychological perspective. Chapter 8 conceptualizes the construction workers' safety performance from safety climate and social exchange perspectives.

Last but not least, the author would like to thank the Australia Government for providing Endeavour Research Fellowship, which supports this book project.

Hong Kong, China and Adelaide, Australia — Rita Yi Man Li

Contents

Chapter 1
Sustainable Construction Waste Management in Australia: A Motivation Perspective

Rita Yi Man Li and Huibin Du

Abstract Construction industry is one of the industries that generates and dumps heaps of waste to landfill. In Australia, about 40 % of the construction and demolition waste is disposed to landfill. Hence, state governments in Australia have initiated various strategies which aim to reduce construction waste. These include levy on construction waste and various proposals of zero waste strategies. Hence, some of the companies reduce, reuse and recycle the construction waste to reduce the amounts of landfill payment. Nevertheless, is monetary expenditure the only effective motivation factor which lead to a reduction in construction waste? Is there any other motivation factors which sucessfully power waste reduction? There are three major contributions in this chapter: (1) it perceives the idea of sustainable construction waste from whole building cycle perspective instead of construction and demolition stage alone. (2) It reviews the construction companies, contractors, government sectors and materials manufacturers' sustainability issues in Australia's construction industry from economics, social and environmental perspectives. (3) It examines the motivation of these companies in sustainable construction waste management under the lens of positive and negative incentive, goal setting and hierarchy of needs theory.

Keywords Motivation · Sustainable development · Construction waste · Content analysis · Australia

R.Y.M. Li (✉)
Department of Economics and Finance / Sustainable Real Estate Research Center, Hong Kong Shue Yan University, Hong Kong, China
e-mail: ritarec1@yahoo.com.hk

R.Y.M. Li
School of Natural and Built Environments, University of South Australia, Adelaide, Australia

H. Du
College of Management & Economics, Tianjin University, Tianjin, China
e-mail: duhuibin@tju.edu.cn

R.Y.M. Li, *Construction Safety and Waste Management*,
Risk Engineering, DOI 10.1007/978-3-319-12430-8_1

1.1 Introduction

The creation of waste pollutes the environment and creates an economic burden on land. Waste leads to an unnecessary loss of natural resources, such as water and embodied energy. The natural resources depletion can only be avoided or reduced via sustainable consumption and strategic waste management based on resource recovery, waste avoidance and efficient material use with less embodied energy. In essence, we aim to reach the target of zero waste where there will be no waste via various transformation and material recycle (Lehmann et al. 2013). Some state governments in Australia have already realised the urgent need to reduce construction waste. For example, the South Australia government has set up a new government branch Zero Waste SA which aims at reaching the goal of zero waste (Table 1.1).

Among all the industries or economic sectors, construction industry has always been identified as one of the major contributors to environment degradation and pollution (Tam and Tam 2008; Lu and Yuan 2011). It generates lots of waste as evidenced in an expansive body of literature. In Australia, the construction and demolition sector generates 38 % of waste (Australia Bureau of Statistics 2010). In Ireland, over 90 % of non-agricultural waste is construction and demolition waste with more than 7 million metric tons of waste is disposed each year (Laefer and Manke 2008). In China, it produces around 30 % of the world's municipal solid waste. Of which 40 % are generated from construction activities (Lu and Yuan 2010).

The construction waste has led to huge negative environmental externality, i.e. the environmental problems generate by the sector itself is borne by the third party (Li 2012b). Its negative impacts include noise pollution, land deterioration and depletion, consumption of non-renewable natural resources, energy consumption, dust and gas emission (Lu and Yuan 2011). Moreover, the industry generates lots of solid waste every year. The offcuts from construction, unused construction materials such as steel, wood, tile, floor slab etc. always end up in landfill. Hence, it is high time to consider means to manage the construction waste in a sustainable manner. In construction sector, modular prefabrication and digital fabrication, provision of adequate storage space, integrated chain management policies, lightweight construction, improvement in staff training, design for re-use and recycling are some of the examples which reduce the natural resources depletion (Lehmann et al. 2013; Al-Hajj and Hamani 2011; Bossink and Brouwers 1996).

This chapter begins with an analysis on Australia state governments' strategies in waste reduction, followed by a discussion on the idea of zero waste, sustainable construction waste management from building life cycle perspective, motivation theories and an analysis on construction companies' strategies from economic, social and environment perspective and their motivations to work towards sustainable waste management.

Table 1.1 Construction waste policy 2014 in different states in Australia

State	Government department	The name of the document
New South Wales	The NSW Environment Protection Authority (EPA) (2007)	http://www.epa.nsw.gov.au/warr/WARRStrategy2007.htm
	NSW Government (2010)	
	New South Wales (2014)	http://www0.health.nsw.gov.au/aboutus/strategicplanning/strategic-planning.asp
	Roads and Maritime Services (2014)	http://www.rms.nsw.gov.au/index.html
Victoria	Sustainability Victoria (2014)	http://www.sustainability.vic.gov.au
	EPA Victoria (2014)	http://www.epa.vic.gov.au
	VicRoads (2014)	http://webapps.vicroads.vic.gov.au/VRNE/csdspeci.nsf
	Metropolitan Waste Management and Resource Recovery Group (2014)	http://www.mwmg.vic.gov.au/
Queensland	Queensland Government (2014)	http://www.derm.qld.gov.au
South Australia	Government of South Australia (2014)	http://www.zerowaste.sa.gov.au/upload/facts-sheets/construction_demolition_3.pdf
Western Australia	Waste Authority (2014)	http://www.wasteauthority.wa.gov.au/programs/partners/strategic-partnerships/strategic-partner-mba
Tasmania	Environment Protection Authority Tasmania (2014)	http://epa.tas.gov.au/policy
	Taswaste (2014)	http://www.taswaste.com
Australian Capital Territory	ACT Government (2014)	http://www.environment.act.gov.au
="top"	ACT Government (2010)	http://www.environment.act.gov.au/__data/assets/pdf_file/0009/576909/23_Butt_Free_Australia.pdf
Northern Territory	Northern Territory Environment Protection Authority (2014)	http://www.ntepa.nt.gov.au

1.2 Solutions Initiated by the State Governments in Australia to Reduce Waste

In Australia, various state governments have initiated a number of projects and proposed a number of suggestions to the industry with regards to waste reduction, reuse and recycle. For example, Australian Capital Territory's government encourages construction demolition waste on-site reuse (ACT Government 2012). In Western Australia, the Master Builders Association Western Australia partners with the Waste Authority to provide MBAWA members with information and tools to reduce waste and increase recycle. They serve for sub-contractors, commercial builders, home builders, suppliers and consultants in commercial and residential sectors. The construction waste management in projects include:

- Educate contractors on how and where to recycle and reuse by conducting site visit on construction site across the Perth and Peel regions;
- Sponsor a number of awards to provide positive recognition and incentives to industry leaders who commit and initiate long-term cultural change in waste reduction and recycle;
- Develop a waste management guide for subcontractors and builders which help resource recovery with a directory of recycling facilities and a template of waste management plan (Waste Authority 2014).

In South Australia, the Government of South Australia (2014) suggests that contractors can reduce waste by:

- Using prefabricated and standardised building materials such as timber, plasterboard and bricks to reduce offcuts waste and costs for waste management and dumping;
- Including a waste minimisation plan in their contracts;
- Reusing or recycling materials;
- Recovering as much material as possible, for example, using construction materials from the demolished structure.

1.3 The Urgency of Zero Waste: We are All Running Out of Landfills

Many cities produce large amounts of waste every day and we are all facing the issue of insufficient landfill space. In spite of the differences in living density, the problem worldwide is surprisingly similar. In the European Union, legislators from Brussels find that the disposal of garbage in landfill sites is getting expensive and there is an increase in pressure to re-use and recycle. San Francisco and Sydney has enough landfill capacity to last until 2014. Beijing will run out of landfill sites in the coming 3 years. As existing landfills in New Delhi will be full soon, New Delhi opens a new landfill site. In view of all the above examples, the idea of zero waste suggest that by way of various construction management strategies, coupled with the 3Rs, i.e. reuse, recycle and reduce, there should not be any building debris and off cuts end up in landfills. Likewise, manufacturers of the construction materials have to make sure that the composition of materials such as concrete and pipes can be reused or recycled to increase the level of sustainability of their products. To make it easier for architects to specify the materials according to their impact due to waste creation from the production process or material extraction, information about the various materials and components should be made publicly available. After all, there is a growing consensus that waste which includes construction waste on sites during construction and demolition process, should be treated as valuable resource and nutrition. Dumping waste into landfill is a failure to design sustainable and recyclable processes and products (Lehmann 2010).

1.4 Zero Waste: A Legend, Myth, or Achievable Goal?

Waste includes any garbage, trash, abandoned or refuse materials which do not have any economic value or functions to anybody who knows its existence. Consequently, the owner has to discard it to landfill. Therefore, the definition of "waste" varies and depends on different perceptions from different persons (Zaman and Lehmann 2011). A "waste" to us may be a valuable resource to the others, so long as there is an appropriate knowledge and infrastructure exists to realize this inherent value (Lehmann et al. 2013). For example, we may discard a Coca Cola can as waste, however, another can collector may grab it into his bag (Fig. 1.1). From this perspective, waste is a product of resource misallocation (Lehmann et al. 2013), the problems of asymmetric information (different people have different information) (Li 2012a, 2014a) such that the waste owner dispose waste instead of giving it to somebody who needs it. It may also argued as a problem of transaction costs which is proposed by Coase which include the costs of sharing the information (turning waste to resource) and negotiating (Li 2008, 2014b). Sometimes, we may even argue that waste should be replaced by the word "resource" (Lehmann 2010).

The idea of "zero waste" was firstly coined and used by Paul Palmer in 1973 who recovered resources from chemicals. Although there is no consensus on the definition of "zero waste" concepts (Zaman and Lehmann 2011). Zaman and Lehmann (2011) suggest that a zero waste city should recycle 100 % of waste and recover 100 % of all

Fig. 1.1 A lady who is collecting the "rubbish" in Rundle Mall (Adelaide, South Australia)'s garbage bin to her bag (author's *photo*)

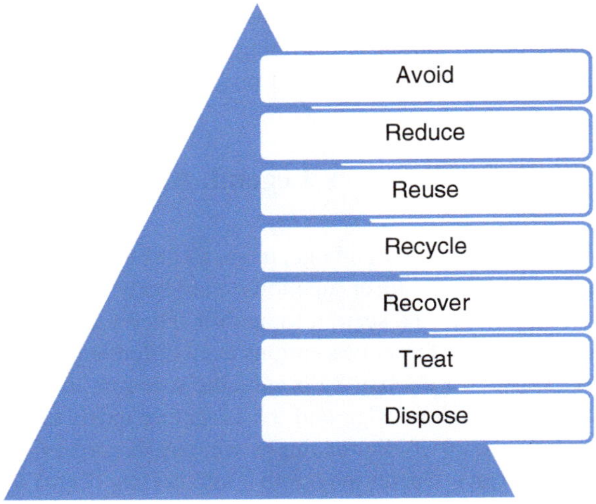

Fig. 1.2 Actions to achieve zero waste (Lehmann et al. 2013)

the resources from waste materials. Zero Waste International Alliance suggests that zero waste is the products and processes design and management which systematically avoid and eliminate the waste of materials so as to conserve and recover all the resources. The concept of zero waste includes a variety of concepts that have been developed for sustainable waste management systems: reduce, reuse, redesign, repair, regenerate, remanufacture, recycle, resell (Fig. 1.2), zero waste dispose to landfill and waste incineration. Thus, zero waste design principle goes beyond recycle which focuses firstly on waste reduction, products reuse and finally recycle and compost the remaining materials. The major barriers to zero waste include the lack of community willingness to pay, lack of consistency in legislation across the states, short term thinking of producers and consumers and the attitude that the cheapest offers get commissioned. The level of waste generation is higher in high-consuming cities as compared to low-consuming cities (Zaman and Lehmann 2011). Nevertheless, there are some barriers which lead to difficulty in realising zero waste. For example:

- Producers and consumers' short term thinking: the costs of dumping to landfill may be cheaper than reuse and recycle;
- Lack of consistency in legislation across different states: the inconsistency in legislation across different states in a county provides loopholes to waste generators to dump waste from one city to another city in another country. For example, when there is landfill charge in city A but not city B, waste generators may transport the waste to city B, leading to the failure in zero waste policies;
- The cheapest contractors' tender may successfully bid for the construction projects. Nevertheless, the relative cheap price may be due to
- Lack of community willingness to pay: as recycle and reuse may incur extra costs, some of the contractors, especially the small and medium size companies may not engage in various construction waste management strategies (Lehmann 2011) (Table 1.2).

Table 1.2 Strategies used to achieve the goal of reduce, reuse, recycle, avoid waste production, treat and dispose the waste in a sustainable manner (Lehmann et al. 2013)

Major types of zero waste strategies in our city	Major actions involved
Awareness, education and research	Zero waste program
	Education
	Zero waste research
Sustainable consumption and behavior	Collaborative consumption, behavior change and sustainable living
New infrastructure and system thinking	New infrastructure, technologies and zero waste governance
Transformed industrial design	Cradle-to-cradle design, cleaner production and producer responsibilities
100 % recycle and recovering	Reduce, repair/reuse, recycle/recovery
Zero depletion, policies and legislation	Zero landfill and incineration legislation, incentives

1.4.1 Zero Waste Policy in Adelaide

After 5 years' work and development, South Australia government has produced the *Draft South Australia's Waste Strategy* 2010–2015 to realise the goal of zero waste. The strategy offers guidelines for South Australia's waste avoidance, recycle and waste. The strategy throws light on two objectives:

(1) To maximize the value of resources, and
(2) To reduce and avoid waste generation (Lehmann 2010).

As the two objectives are interrelated, some actions in the strategy are applicable to these two objectives. They include the newly proposed targets for construction and demolition, industrial, commercial and municipal solid waste. Funded by levies from landfills, Zero Waste SA was established in 2003, it is one of the few government agencies which advocate zero waste at the forefront of waste avoidance in the World (Lehmann 2010).

1.5 Sustainable Construction Waste Management: Building Cycle Perspective

Sustainable development highlights the importance of the co-development of environment, social and economics (Li 2011; Li and Ah Pak 2010) (Fig. 1.3). The idea rests on the notion that the activities of our present generation should

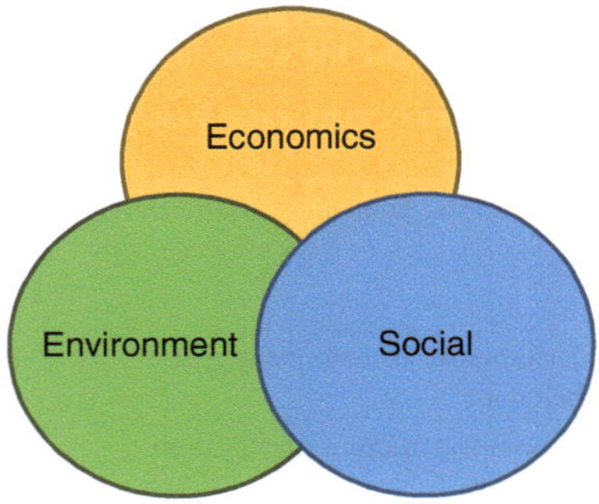

Fig. 1.3 The three major elements in sustainable development (Li 2011)

not adversely affect the future generations. From this perspective, sustainable construction waste management should not focus on environmental perspective alone but to strike a balance between economics, social and environment. Although many of the research sheds light on construction waste management, they mainly focus on the construction process. Nevertheless, we should view the issue from the whole building life cycle perspective (Tables 1.3, 1.4 and 1.5).

1. Wrong inception will lead to unnecessary demolition during construction stage;
2. Poor design which lack of future foresight will lead to premature demolition/renovation in occupancy stage;
3. Poor construction waste management, for example, insufficient communications between workers may lead to excessive offcuts during construction phrase;
4. Poor building usage will bring an early end to the building (for example, a building was collapsed due to retrofit workers accidentally demolish part of the building structure in Hong Kong is one of the vivid example);
5. Poor demolition plan and work may lead to insufficient waste to be reused in the future. For example some of the used glass may be recycled; part of the building materials can be used in the newly constructed building.

In the same vein, the document prepared by Edge Environment Pty Ltd (2012) suggested that sustainable construction waste management should not solely focus on the construction or demolition stage. The ideal construction waste management should encompass the whole building life cycle which includes project inception, design, construction, operation and demolition stage. Details of how the construction waste management from whole building life cycle perspective are illustrated in Table 1.5.

Table 1.3 Economic considerations in building life (Edge Environment Pty Ltd 2012)

Project inception stage	
Supply and demand	Local, regional, national, and global's market supply and demand of similar products/projects at present and in the future
Market forecast	Market size, price, marketing strategies, and marketing targets
Scale and business scope	Project scale and the business scope which affect project profitability
Effects on local economy	A project should take advantage of the infrastructure in the local economy to generate economic benefits and serve the local economy in return
Life cycle cost analysis	The total cost for building-up, operating, maintaining, and disposing a construction project over its life
Life cycle profit analysis	The total profit arise from project life cycle
Capital budget	Capital budget should include project total cost planning and controlling
Finance plan	Project finance schedule should be included
Investment plan	Arrangement of a cash plan, liquid and fixed capital for investment at project inception stage
Employment	The project should provide local employment opportunities
Project design stage	
Consideration of life cycle cost	Consider the total cost involved in project life cycle, including site formation, construction, operation, maintenance cost and demolition cost
Project inception stage	
Materials choice	Economy, durability and availability for material selection
Project construction stage	
Loan interests	The interests for capital cost paid for liquid capital and a fixed loan
Opportunity cost	Liquid and fixed capital for the project will lead to the lose in opportunities in other projects
Labor cost	Salaries paid to general construction workers, stonemasons, pipe-liners etc.
Professional fees	Fees that are paid to various professionals and consultants such as environmental, engineers, ecological, legal and geological experts
Materials cost	Costs for all types of materials such as lime, concrete, timber, steel and bamboo
Energy cost	Costs for various types of energy such as oil, electricity, coal and gas
Water cost	Costs for water resources such as surface water, and ground water
Equipment cost	Costs for various tools, tower cranes and vehicles
Equipment purchase cost	Costs for various equipment such as escalators, elevators, HVAC systems and plants

(continued)

Table 1.3 (continued)

Installation cost	Costs for all kinds of equipment and facilities installation
Direct employment	Provisions of job opportunities from implementing project to local labour market, such as construction workers, engineers and professionals
Indirect employment	The up-and-down stream industries and services to construction employment
Project operation stage	
Distribution of project income	Paybacks, dividends and reinvestment
Balance sheet from project operation	Develop a balance sheet to check with the project time and cost
Labour cost	Salaries for engineers, workers, managerial staff and professionals
General expenses	Daily consumables, water, gas and electricity
Materials cost	Various materials for project maintenance and operation
Logistics costs	Materials stock costs, transportation and procurement
Marketing costs	Resource investment for promotion, advertising and market analysis
Training costs	Training staff for human resources quality improvement
Local economic environment improvement	The project should benefit the local society economically
Direct job opportunities in project operation	Employment costs for managers, professionals and workers
Indirect job opportunities	Job opportunities associated with along up-and-down stream industries during project operation
Demolition stage	
Labour cost	Planning, managing and operating project demolition's human resources
Energy for demolition procedures	Cutting, transferring and moving the materials produced from demolition site
Costs for waste disposal	Loading and unloading waste, transportation and charges for disposals
Compensation	Affected stakeholders' compensation during demolition
Deployment or dissolution of project staff	Provision of pensions, unemployment compensation
Compensation paid for the polluted environment	Compensation paid to the local residents for the damaged environment land, water, and ecosystems
Land redevelopment's value	The land value after demolition for re-development
Residual value	Valuable residues, such as brick, glass, equipment, steel, timber for recycle and reuse
Land for new development	Provision of land upon project demolition completion to enable new projects' implementation according to the demand of local community
Employment opportunity	Provision of employment opportunity during demolition for site work, transportation and disposal

Table 1.4 Environmental considerations in building life cycle

Environment	
Project inception stage	
Land use	The land selection for the project should protect natural resources and cropland
Eco-environment sensitivity	Avoiding as much as possible the adverse impacts on surroundings near the project
Ecological assessment	Examining potential ecological benefits and risks associated with the proposed project
Air assessment	Access the potential air pollution from the proposed project and the impact on local climate
Water assessment	Examine the potential water pollution from the proposed project, including project's consumption on water resources, surface and ground water
Noise assessment	Access the potential noise pollution during construction and operation
Waste assessment	Examine waste generation in construction and operation
Project design stage	
Designer	Knowledge in energy savings and environmental issues
Environmental design	Incorporate all the environmental considerations into project design for demolition, construction, recycling, disposal and operation
Modular and standardized design	Use of standardized and modular components to enhance buildability such that waste generation can be reduced
Land use pollution	Utilize the land effectively to avoid land pollution
Natural habitat destruction	Protect the environment for animals and human
Air emission and pollution	Generation of SO_2, NO_2, NO, CO_2, CO
Noise pollution	Vibration and noise induced from the project
Water pollution and discharges	Release of organic pollutants and chemical waste to water ways
Waste generation	Waste produced from the project
Comfort disturbance	Impact on the balance of ecosystem and people's living environment
Energy and resource consumption	Resources consumption and energy saving which includes electricity and water
Using renewable materials	Using typical renewable and reproducible materials such as wheat straw cabinetry, cork, fast-growing poplar, bamboo
Ozone protection	Reduce the release of hydro-chlorofluorocarbons and chlorofluoro-carbons and hence protect the Ozone layer
Off-site prefabrication	Reduce on-site construction waste
Material reuse	Reuse of building components such as earth, rubble, steel, timber and concrete
Structural operations	Consideration being given to the reduction of earthwork and excavation, formwork, reinforcement, concreting and waste treatment during structural operation
Internal and external operations	Control environmental impacts from waste treatment, roofing, walling, painting, insulation, plumbing and drainage, landscaping and component installation

(continued)

Table 1.4 (continued)

Environment	
Project organization	Environmental management task force, resource coordination, supervision and cooperation culture
Environmental resource management	Resource inputs for implementation of environmental management such as plant, labour, finance and material
Organizational policy	Establishment of environment management system, application of environmental management standards such as ISO 14000, project manuals, programs, progress control reports
Environmental management information communication	Managing project environment information via information management facilities and management expertise
Environmental management technology	Environmental experts and management facilities, resource and energy saving technology, pollution and waste reduction technology
Environmental regulations	Environmental protection regulations and law on construction activities
Project operation stage	
Land contamination	Release of chemical wastes via landfills and dumping
Air pollution	Air pollutants generation such as NO_2, NO, SO, CO_2 and CO
Water pollution	Organic pollutants and chemical wastes release to water ways
Noise pollution	Vibration and noise generated by occupants
Waste generation	Wastes produced by occupants
Ecological impacts	Negative impacts generated by occupants to flora, ecosystems and fauna
Various energy consumption	Electrical, lighting and other energy appliances' energy consumption
Water consumption	Water used for heating, hygiene and cooling
Raw material consumption	Use of non-renewable and renewable raw materials
Staff's environmental awareness training	Provide training and education programs to different levels of employees
Environment friendly facilities management	Improve facilities' productivity, reduce pollutant generation and resource consumption
Demolition stage	
Demolition plan	Plan on waste and hazard materials' recycle or reduction
Demolition control	Control and supervision on the demolition activities to protect the environment
Environmental friendly demolition	Adoption of technologies to maximize reusing and recycling waste, alleviate the adverse impact on eco-environment systems and neighbourhood
Environmental information and policy Communication	Knowledge about legislations, regulations, environmental techniques and environmental policies
Waste classification	Good classification of demolition wastes for effective disposal and treatment
Special waste treatment	Special treatment given to radioactive chemicals, heavy metals, toxic materials that are released from demolition
Waste reuse and recycle	Recycling and reclaiming the useful materials such as timber, steel, glass, brick and some equipment

Table 1.5 Social considerations in building life cycle

Social	
Project inception Stage	
Land use	Considering that the land selection for project site should protect cropland and natural resources
Conversion of cultural and natural heritage	Reduce negative impacts from project development on any cultural heritage
Infrastructure capacity-building	The project improves local infrastructure capacity, such as power, drainage, sewage, road, transportation, dinning, communication, recreation, education, shopping, financing and medical
Community amenities	Any future safety risks to the project users and public should be identified
Project design stage	
Project layout	Standard dimension as required in design specifications
Safety design	Design for emergencies such as fire, earthquake, eco-environmental accidents and flood
Security consideration	Security alarm and security screen installation
Life cycle design	Effective communications among government officers, clients, designers, environmental professionals to ensure that all the environmental requirements are incorporated into design
Construction stage	
Site security	Various site safety measures
Construction safety	Safety facilities, measures and insurance for staff
Public safety	Provision of signal systems, facilities, warning boards and safety measures for the general public
Improvement of infrastructure	Provision of better sewage, drainage, heating, message, electrical systems and road
Infrastructure burden	Extra demand for road, water, services, space and energy for the project
Health and safety risks	Ensure on-site safety and health by providing on-site training programs to employees and supervision so as to reduce the number of accidents
Health and safety	Provision of safety and health care safety facilities, emphasize on-site hygiene
Project operation stage	
Provision of services	Benefits of local communities' living standard improvement
Provision of facilities	Provision of facilities and space which benefits local's development
Demolition	
Operational safety	Labors and the public's safety risks during project demolition from dismantling, explosion, radioactive and toxic materials
Communication to the general public	Enhance general public's awareness on project demolition's impacts

1.6 Motivations for Sustainable Zero Waste Construction Waste Management

1.6.1 Incentive

Positive incentive motivation encompasses positive mood valence, increase energy and activation. Such states of positive activation are characterized by feelings of desire, enthusiasm, want, excitement, achievement, energy, elation and potency (Morrone et al. 2000). In general, these incentives may be distributed to groups or individuals (Forno and Merlone 2010). Previous literature evidences that many of negative incentive often related to extra costs (in terms of monetary and non-monetary expenditures. For example, Guo and Sun (2014) suggest to adopt congestion pricing. In the same vein, Theory X suggests that men have to be directed, forced and threatened with penalty for achieving certain organizational objectives (Li 2009; Li and Poon 2009) (Table 1.6).

Previous construction waste research shows that there are many different types of punishment and rewards which enhance workers' awareness of the financial and environmental aspects of 3Rs, i.e. reuse, recycle and reduce. In terms of positive strategies, a few scholars suggest that reward schemes successfully motivate construction workers in waste reduction. Financial benefits can be gained by adopting waste-minimization strategies to reduce landfill levy and material costs. Special reward programs are found to be effective and have been used as a performance-dependent monetary reward system in the construction industry. Some organizations penalise their subcontractors by withholding payment if the environmental performance is unsatisfactory (Tam and Tam 2008).

1.6.2 Goal Setting Theory

Goal-setting theory suggests that goals serve as a reference standard in cognitive comparison process of self-satisfaction and evaluation (Hsiaw 2013). It suggests that goals serve as a standard in cognitive comparison process of self-satisfaction and fullfilment (Hsiaw 2013). The current pool of literature suggests that, to increase the likelihood of successful achievement in some goals so as to achieve the optimal level of performance, these goals should be:

(1) The level of difficulty is challenging but can be achieved (Edward and Clenney 2012);
(2) They are specific (Smith et al. 2013).

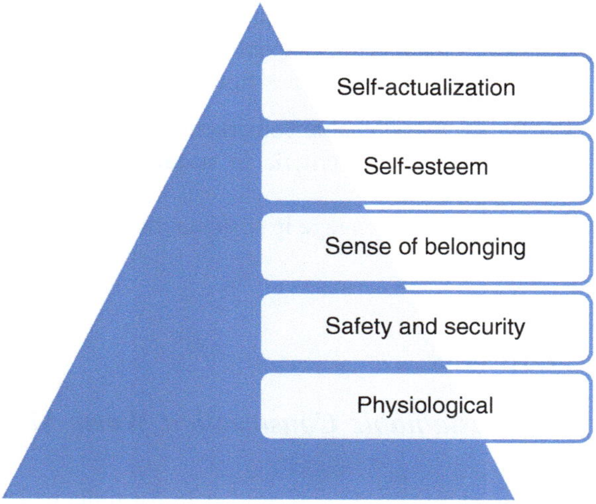

Fig. 1.4 The hierarchy of needs theory (Gambrel and Cianci 2003)

1.6.3 Hierarchy of Needs Theory

The validity of Hierarchy of needs theory has been highlighted in Noltemeyer et al. (2012) which is quoted from Maslow's paper (1943) "[i]t is quite true that man lives by bread alone—when there is no bread. But what happens to man's desires when there is plenty of bread and when his belly is chronically filled? At once other (and "higher") needs emerge and these, rather than physiological hungers, dominate the organism. And when these in turn are satisfied, again new (and still "higher") needs emerge and so on. This is what we mean by saying that the basic human needs are organised into a hierarchy of relative prepotency" (Fig. 1.4).

As Hierarchy of needs theory accurately describes human motivation, it has always been identified as one of the most powerful instrument to explain why somebody will do something or are propelled to do something. Nevertheless, the lack of empirical testing, vagueness of the concept have become some of the major criticisms of the theory (Saeednia and Nor 2013).

1.7 Research Method

The research method for this chapter is content analysis. The first content analysis was used in the 1950s and was predominately quantitative at that time. Later, the technique includes the interpretations of latent content where similar information is categorised into different groups according to the meaning of the content to create systematic and objective criteria for analysing symbolic content of the documents (Li 2013). Kassarjian (1977) posits that it is a process of information processing where written communications is transformed via systematic and objective rules of categorisation, such that we can summarise the data for comparison. In this chapter, we categorize the 15 case studies from Edge Environment Pty Ltd (2012) according to

(1) Types of sustainable construction management activities;
(2) The three major criteria in sustainable development (economics, social and environment);
(3) Motivation to engage in sustainable construction waste management activities.

1.8 Results

1.8.1 Sustainable Construction Waste Management

Table 1.6 shows the 15 case studies in Australia which studies the possibility for the co-development in social, economic and environmental aspect in construction projects and materials production companies which aim to reduce construction waste. It is found that quite a number of these case studies can attain the win-win-win economic, social and environment development, e.g. Ontera Modular Carpets, Fairfield City Council Sustainable Resource Centre, Fletcher and CSR Gyprock™ etc. (Edge Environment Pty Ltd 2012).

From economic perspective, sustainable zero waste activity does not necessary leads to high costs. Some of them evidence that the sustainable action may even lead to higher economic profit as home-buyers prefer to purchase residential units from builders who use products with low environmental impact and recycle as reported in CSR Gyprock™—a gypsum board take-back scheme. From social perspective, it may enhance public health and safety. For example, PIPA recycling scheme recycles PVC, reduces PVC in landfill site so as to lower the health risks of the potential PVC fire in landfill. Some of the companies find that engaging sustainable building activities can achieve corporate social responsibility (CSR). For example, Fletcher finds that when the company turns the used glass to insulation materials, it achieves the corporate social responsibility to produce a product from a high proportion of post-consumer materials. Built Environs'100 Hutt Street project in Adelaide also considers sustainable construction waste management as a kind of Corporate Social Responsibility. From environment perspective, it helps reduce landfill pressure, e.g. PIPA recycles PVC pipe, Waverley Council recycle glass recycle timber, the Hazelmere Timber Recycling Centre and James Hardie Industrial Ecology Initiative recycle timber, Ontera carpet re-collects and reuses the used carpet (Edge Environment Pty Ltd 2012). Some of them, however, fail to benefit from economic saving. For example, the Old Leura Dairy Buildings project reports an increase in construction costs (Edge Environment Pty Ltd 2012).

Table 1.6 Case studies of construction companies which engage in sustainable waste management (author's summary)

Name of the project	Details of the project	Economics	Environment	Social	Motivation
Fairfield city council sustainable resource centre	95 % of concrete and 65 % of asphalt is recycled	Send the waste to landfill will incur a cost to the council, as fewer materials will be sent to landfill, costs can be reduced	Divert over 100,000 tonnes of material from landfill per year and save the valuable land	Recognition—the Fairfield City Council won the local government award "Best Specific Environmental Initiative"	Turn expense in the balance sheet to income
		The new recycle products are produced			Save landfill site and award
Ontera modular carpets	Re-collect and reuse around 600 tonnes of the used ontera carpet	Generate economic benefit	Lower 25 % of the carpet waste	Enhance staff engagement	ISO 14001 certified
			Lower more than 19 % carbon dioxide emissions		
			Increase recycled content by 50 %		
			Save around 2,000 litres water per month in dying process		
		Reduce landfill and utility charges	Reduce energy	Increase their loyalty to the company	The carpet has been specified in Australia's Green Building council's green star rating tool
Armstrong Australia	Collect end-of-life and off-cuts flooring to make new PVC flooring	Create a new market for PVC waste	Reduce PVC toxicity from disposal, carbon dioxide emissions and pressure on landfill due to disposal	Not mentioned	Product stewardship program (a voluntary program which works for end-of-life product)
					Demonstrate good waste management
					Reduce landfill charges

(continued)

Table 1.6 (continued)

Name of the project	Details of the project	Economics	Environment	Social	Motivation
Fletcher	It uses used glass from the waste stream to make glass wool insulation	Post-consumer waste glass is cheaper than virgin glass material	The usage of recycled glass reduces the energy used in the furnace	Fletcher's corporate social responsibility: to produce a product from a high proportion of post-consumer material	Using post-consumer waste glass is cheaper than virgin glass material
		It reduces the energy for manufacturing glass wool insulation			It reduces the process energy
CSR gyprock™—a gypsum board take-back scheme	Upon the completion of gypsum board installation, all offcut plasterboard material to be recycled are kept for collection	Home-buyers prefer to buy houses from builders who use products with low environmental impact and recycle	Decrease in raw material use	Better on-site resource management, produces safer and cleaner sites	Corporate social responsibility, reduce raw materials and disposal to landfill, business offering improvement
				Government officials are increasingly insisting on better waste management	
	The contractor arranges collection with CSR Gyprock™'s recycling contractor who guaranteed the material will be 100 % recycled upon some fees charges	Reduce the cost of landfill fees and site clean-up	Reduce disposal to landfill	It achieves corporate social responsibility	

(continued)

Table 1.6 (continued)

Name of the project	Details of the project	Economics	Environment	Social	Motivation
The plastic industry pipe association (PIPA) recycling scheme recycles	300 to 400 tonnes of demolition and offcut PVC pipes are recycled	Save materials	Reduce waste from landfill and commensurate dioxin	Reduce PVC in landfill so as to lower the health risks of the possible PVC landfill fire	When PVC is burnt and reach the end of its life, uncontrolled landfill fires may occur
					Legal rules and the industry's self-regulation such as the Green Building Council of Australia's PVC credits, meet the best practice and other design guidelines which specify PVC products' standards
					The industry strives for best practice
			The total life cycle of PVC pipe is more sustainable		Increase in recycling rates can save virgin material
James Hardie industrial ecology initiative	Recycle a portion of by-product to manufacture pallets	Maximize efficiency and replace valuable natural resources with by-products	Reduce the demand for energy and natural resources	Reduce the landfill pressure	To achieve the firm's corporate social responsibility
	Reduce the demand for timber and replace timber on pallets	Partner with Rosehill plant—a major producer of road base materials where by-product fines are reduced	Reduce the demand for quarry materials and forestry resources which include sand, limestone, timber and other products extract from mines		Cement manufacturers' initiatives to look for alternative materials for cement's manufacturing and process improvement
	Reduce landfill pressure by more than 80 % (16,000 tonnes of waste) every year				

(continued)

Table 1.6 (continued)

Name of the project	Details of the project	Economics	Environment	Social	Motivation
The hazelmere timber recycling centre (Hazelmere) in perth	Recycle 10,000 tonnes of timber per year	Save the community and councils' disposal fees of around $50 per tonnes which is less than landfill costs	To reduce landfill pressure	Not mentioned	The Western Australia Government Waste Authority's "towards Zero Waste" and reduce the landfill pressure
			Timber is processed to reusable woodchip substantially reduce tree cutting		
Kennedy's timber in queensland recycles power poles into valuable hardwood products.	Kennedy Timber recycles old Energex power poles and then sells the recycled timer	Redundant power poles impose a carbon cost to the company	Reduce carbon emission and landfill pressure	To develop the new standard for the use of recycled timber	To recover the value of the untreated parts of the power pole timber
		Eliminate carbon emissions			
		Reduce the costs and turn the liability into an asset			
The old Leura dairy the OLD buildings	The building is a development of six luxury eco-friendly buildings built by the Hennessey Family	The costs of labour for recycle works are equal to or even more than the cost of using new or virgin materials	Framing timbers from a local recyclers. Yard lining, decking boards and weather boards, railway bridge timbers, floorboards from old woolsheds, corrugated iron sheets from roofing is used for cladding sinks, doors light fittings, toilets and windows	Tradesmen gain an experience of reclaimed materials	To create interesting buildings with character which attracts people, create a place with unique characteristics
	The buildings use up to 95 % (in terms of costs) locally recycled and sourced reclaimed materials		40 tonnes of bricks are made from the previous Katoomba Ice Works	Knowledge of the new methods of construction increases	

(continued)

Table 1.6 (continued)

Name of the project	Details of the project	Economics	Environment	Social	Motivation
	Provide corporate retreat and tourist accommodation in Blue Mountains (NSW)		The carpet is used as additional insulation beneath the floor	New methods have already been tried, tested, save the product development time in the future	The key motivation is a philosophical commitment to ecological sustainability
Waverley council	Recycled glass in pavement construction	If concrete industry uses the unused crushed glass fines, 75,000 tonnes of natural sand at $30 per tonnes can save them $2.25 million	Reduce landfill pressure	Not mentioned	60,000 tonnes in Sydney and around 75,000 tonnes of crushed glass fines in New South Wales are destined in landfill unless an alternative use can be found
					If the concrete industry uses the unused crushed glass fines, 75,000 tonnes of natural sand at $30 per tonne would save them $2.25 million
					By enaging stakeholders to recycle, reporting the tonnes of recycled glass material used and demonstrating pavement construction by glass, it helps achieve the sustainability targets

(continued)

Table 1.6 (continued)

Name of the project	Details of the project	Economics	Environment	Social	Motivation
Trevor pearcey house	Trevor Pearcey House, a 19-year-old building in the Fern Hill Technology Park, is refurbished to become the new head office for Australian Ethical Investment	The material cost savings from the carpet can be spent on other area of the sustainable design	Reduce ducted skirting, power point and switch face plates, partition wall studs, electrical wiring, plaster-board and frames for doors, door handles, door stops, internal glass blocks, windows	A recognition for innovative ways of re-used and recycled materials in buildings	Business philosophy in sustainability. A target of six stars in GBCA's Green Star rating too
			Reuse carpet tiles and supplemented with recycled carpet Ontera Carpets		
	It commits to recycle and re-use materials from the demolition phase of the project		Reproduce bike enclosure from steel hanging frames and mesh in ceiling		
	A recycle rate of more than 80 percent by weight		90 % of the joinery cupboards made from old cupboards are used in walls and floors		
	Innovative methods incorporate recycle and reuse materials into the new building		Recycling rate exceeds 80 % by weight		The Australian Capital Territory Government's No Waste by 2010 policy

(continued)

Table 1.6 (continued)

Name of the project	Details of the project	Economics	Environment	Social	Motivation
Built environs: 100 hutt street, Adelaide	Recycle/reuse 95.1 percent of construction waste by weight from construction activities	Create job opportunities	Reduce landfill and waste by conserving resources	Increase knowledge and experience on waste management	It is a flagship project to demonstrate what their company could do
			Reducing pollution and energy conservation (turning recycled material into new material takes less energy than from scratch)		
			Reuse 95.1 % by weight of construction waste which include black caesar stone removed from a prominent South Australian Public building (part of a previous refurbishment project)		
			Recycle mechanical spiral ductwork is reused for 295 pot plant sleeves		
			Recycled timber is used for wall noggins for partition walls		
			Recycled fire sprinkler pipework is used for PPE display		The experience can be used in other projects
			Timber palettes used to store and deliver mechanical ductwork, to create partitions in "palette room".		Pursuing the project philosophy of sustainable methodologies and environmental design

(continued)

Table 1.6 (continued)

Name of the project	Details of the project	Economics	Environment	Social	Motivation
			Surplus concrete reinforcing mesh was then powder coated and used in part of the reception area		
			Reuse 40-gallon steel drums salvaged from Built Environs' plant yard for seating in break-out area	There is flow-on impact on staff's commitment to manage waste effectively	The team ensured waste materials would be re-used within the fit-out as an alternative to conventional materials is a corporate social responsibility
		The experience and knowledge in the project will generate economic benefit as clients often request for a repeat in the company's previous good work	Reused wire mesh from surplus stock on previous projects is used for stair balustrade design	The project has instilled waste minimization as a core culture	They wish to increase the knowledge for visiting subcontractors, future project consultancy teams, their employees and general visitors
Fairfield City council: a sustainable community building	95 % recycled concrete load-bearing foundation slab has never been tried before anywhere else in the world	N/A	Convert waste straw to valuable building resources	10 times better insulation than double-brick cavity wall	To keep its promise on the commitment to sustainability, Cities for Climate Protection and Local Agenda 21
			It reduces greenhouse gas generation (straw is often burned)		
			It enhances the efficiency of solar energy design walls which are 30 times less energy intensive than wooden frame walls		To inspire the community to make more environmental sustainable choices as routines

(continued)

Table 1.6 (continued)

Name of the project	Details of the project	Economics	Environment	Social	Motivation
Pod scrap bag program	Scrap bags are supplied with expanded polystyrene (EPS) pod deliveries to construction sites to aid the separation of EPS offcuts	Reduce virgin material required to make pods	All scrap and offcuts can be reused and recycled in new pods	N/A	The persistent waste on and around construction sites from the use of waffle pods pose threat to regulate its use
	The filled scrap bags are collected and taken back to EPS manufacturer. After that, they are granulated and recycled as new waffle pods and other construction products				

1.8.2 *Construction Waste Reduction Should Be Building Life Cycle Concept and Action*

The research results also show that waste reduction can be done in various stages of the building life cycle which includes construction, occupation and demolition phrases. For example, in construction demolition and retrofit stage, Fletcher uses used glass from the waste stream to make glass wool insulation. In construction stage, The Vinyl Council of Australia works with Armstrong Australia Collect end-of-life and off-cuts flooring to make new PVC flooring. Fairfield City Council used 95 % recycled concrete load-bearing foundation slab a sustainable community building In occupation stage, Ontera Modular Carpet recollected the old and used carpet for reuse (Edge Environment Pty Ltd 2012). In building refurbishment, 80 % of the building materials are reused or recycled in Trevor Pearcey House project.

1.8.3 *Motivations of Sustainable and Zero Waste Management*

1.8.3.1 Incentive Theory Perspective

Many companies are of the view that the landfill costs are one of the major motivation which motivates them to reduce waste via certain methods. On top of the land fill charged by the operators, an additional charge levied by the state and territory jurisdictions may also be possible. In New South Wales, the government's Section 88 Landfill Levy charges are applicable to regulated area. They motivate the industry manufacturers and contractors to re-use and recycle more materials in the construction industry. Moreover, as making new construction materials need extra energy and given the rising trend of electricity prices, e.g. AGL in Adelaide has already announced the increase in the electricity rate again since August 2014. The rising costs will directly affect the building materials and products costs and reflected in the price of construction products and materials. These, again, will motivate more reuse and recycle activities on sites so as to reduce the embodied energy of production (Edge Environment Pty Ltd 2012).

Example of positive incentive in case studies which successfully motivate the companies engage in construction waste management:

- Fletcher uses post-consumer waste glass to make glass wool insulation as it is cheaper than virgin glass material (Edge Environment Pty Ltd 2012).

Example of negative incentive in case studies which successfully motivate the companies engage in construction waste management:

- PIPA recycle PVC because of regulation and the industry's requirement specified in the Green Building Council of Australia's PVC credits and design guidelines which specify PVC products' standards (Edge Environment Pty Ltd 2012).

1.8.3.2 Goal Setting Perspective

The companies that were covered in these case studies coincide the idea suggested by Hsiaw (2013) that construction companies are motivated by the goals' self-satisfaction and fulfilment. For example:

- Australian Ethical Investment reuse as high proportion of the materials used as possible to reach the goal of receiving six stars in GBCA's Green Star rating tool.
- As Ontera Modular Carpets sets a goal to receive ISO 14001 so it collects and reuses tonnes of used carpet (Edge Environment Pty Ltd 2012).

1.8.3.3 Hierarchy of Needs Theory: Waste Management Is a Kind of Self-actualisation Activities

The study reviews that quite a large number of the companies act sustainably due to corporate social responsibility. This is a kind of higher level motivation as stated in Hierarchy of Needs' theory. For example:

CSR Gyprock™—a gypsum board take-back scheme is implemented because the company feel that it is a kind of corporate social responsibility to reduce raw materials and waste disposal to landfill.

James Hardie Industrial Ecology Initiative reduce timber and reuse some of the by-products during production.

Built Environs's project in Adelaide ensured waste materials would be re-used in the fit-out work as an alternative to conventional materials (Edge Environment Pty Ltd 2012).

1.9 Conclusion

As construction waste accounts for a large proportion of waste in many different countries, leads to pollution and environmental externalities, many state governments in Australia initiate different strategies to reduce construction waste. The South Australia government, for example, proposes zero waste strategies. Prefabrication, standard sized construction materials are proposed by construction industry to reduce waste. This chapter suggests that sustainable construction waste management, i.e. co-development in economic, social and environment, was feasible in many case studies despite a few experienced an increase in construction costs. Furthermore, these construction companies were not solely motivated by monetary incentives, such as landfill levy. Quite a number of them revealed that they were motivated by corporate social responsibility.

References

ACT Government. (2010). ACT sustainable waste draft strategy 2010–2025. Retrieved August 5, 2014 from http://www.environment.act.gov.au/__data/assets/pdf_file/0009/576909/23_Butt_Free_Australia.pdf.

ACT Government. (2012). ACT waste management strategy. Retrieved September 6, 2014 from http://www.environment.act.gov.au/__data/assets/pdf_file/0007/576916/EDS_ACT_Waste_Strategy_Policy_23AUG2012_Web.pdf.

ACT Government. (2014). Environment and planning directorate. Retrieved August 5, 2014 from http://www.environment.act.gov.au.

Al-Hajj, A., & Hamani, K. (2011). Material waste in the UAE construction Industry: Main causes and minimization practices. *Architectural Engineering & Design Management, 7*, 221–235.

Australia Bureau of Statistics. (2010). 4613.0—Australia's environment: Issues and trends, Jan 2010. Retrieved September 7, 2014 from .

Bossink, B. G., & Brouwers, H. H. (1996). Construction waste: Quantification and source evaluation. *Journal of Construction Engineering & Management, 122*, 55.

Edge Environment PTY LTD. (2012). Construction and demolition waste guide—recycling and re-use across the supply chain.

Edward, W., & Clenney, E. F. (2012). Extremely difficult negotiator goals: Do they follow the predictions of goal-setting theory? *Organizational Behavior and Human Decision Processes, 118*, 108–115.

Environment Protection Authority Tasmania. (2014). About policy and legislation. Retrieved August 7, 2014 from http://epa.tas.gov.au/policy.

EPA Victoria. (2014). EPA victoria. Retrieved August 5, 2014 from http://www.epa.vic.gov.au.

Forno, A. D., & Merlone, U. (2010). Incentives and individual motivation in supervised work groups. *European Journal of Operational Research, 207*, 878–885.

Gambrel, P. A., & Cianci, R. (2003). Maslow's hierarchy of needs: Does it apply in a collectivist culture. *Journal of Applied Management and Entrepreneurship, 8*, 143–161.

Government of South Australia. (2014). Construction and demolition materials. Retrieved September 6, 2014 from http://www.zerowaste.sa.gov.au/upload/facts-sheets/construction_demolition_3.pdf.

Guo, X., & Sun, H. (2014). Reward and coarse toll in competitive system with two modes. *Procedia—Social and Behavioral Sciences, 138*, 205–213.

Hsiaw, A. (2013). Goal-setting and self-control. *Journal of Economic Theory, 148*, 601–626.

Kassarjian, H. H. (1977). Content analysis in consumer research. *The Journal of Consumer Research, 4*, 8–18.

Laefer, D. F., & Manke, J. P. (2008). Building reuse assessment for sustainable urban reconstruction. *Journal of Construction Engineering & Management, 134*, 217–227.

Lehmann, S. (2010). Resource recovery and materials flow in the city: Zero waste and sustainable consumption as paradigms in urban development. *Sustainable Development Law & Policy, 11*, 28–68.

Lehmann, S. (2011). Optimizing urban material flows and waste streams in urban development through principles of zero waste and sustainable consumption. *Sustainability, 3*, 155–183.

Lehmann, S., Zaman, A. U., Devlin, J., & Holyoak, N. (2013). Supporting urban planning of low-carbon precincts: Integrated demand forecasting. *Sustainability, 5*, 5289–5318.

Li, R. Y. M. (2008). Nature of the firm: A study on developers in China and Hong Kong. *Journal of Applied Economic Sciences, 3*, 479–486.

Li, R. Y. M. (2009). Achieving compliance with environmental health-related land use planning conditions in Hong Kong: Perspectives from traditional motivation theories. *Journal of Environmental Health, 71*(2), 177–188.

Li, R. Y. M. (2011). *Building our sustainable cities*. Illinois: Common Ground Publishing.

Li, R. Y. M. (2012a). Econometric modeling of risk adverse behaviours of entrepreneurs in the provision of house fittings in China. *Australasian Journal of Construction Economics and Building, 12*, 72–82.
Li, R. Y. M. (2012b). The internalization of environmental externalities affecting dwellings: A review of court cases in Hong Kong. *Economic Affairs, 32*, 81–87.
Li, R. Y. M. (2013). The usage of automation system in smart home to provide a sustainable indoor environment: A content analysis in Web 1.0. *International Journal of Smart Home, 7*, 47–59.
Li, R. Y. M. (2014a). *Law, economics and finance of the real estate market—a perspective of Hong Kong and Singapore*. New York: Springer.
Li, R. Y. M. (2014b). Transaction costs, firms' growth and oligopoly: Case studies in Hong Kong real estate agencies' branch locations. *Asian Social Science, 10*, 40–52.
Li, R. Y. M., & Ah Pak, D. (2010). Resistance and Motivation to share sustainable development knowledge by web 2.0. *Journal of Information & Knowledge Management, 9*, 251–262.
Li, R. Y. M., & Poon, S. W. (2009). Future motivation in construction safety knowledge sharing by means of information technology in Hong Kong. *Journal of Applied Economic Sciences, 4*, 457–472.
Lu, W., & Yuan, H. (2010). Exploring critical success factors for waste management in construction projects of China. *Resources, Conservation and Recycling, 55*, 201–208.
Lu, W., & Yuan, H. (2011). A framework for understanding waste management studies in construction. *Waste Management, 31*, 1252–1260.
Maslow, A. H. (1943). A theory of human motivation. *Psychological Review, 50*, 370–396.
Metropolitan Waste and Resource Recovery Group. (2014). Enabling change for a sustainable future. Retrieved August 5, 2014 from http://www.mwmg.vic.gov.au/.
Morrone, J. V., Depue, R. A., Scherer, A. J., & White, T. L. (2000). Film-induced incentive motivation and positive activation in relation to agentic and affliative components of extraversion. *Personality and Individual Differences, 29*, 199–216.
New South Wales. (2014). Strategic planning framework. Retrieved August 1, 2014 from http://www0.health.nsw.gov.au/aboutus/strategicplanning/strategic-planning.asp.
Noltemeyer, A., Bush, K., Patton, J., & Bergen, D. (2012). The relationship among deficiency needs and growth needs: An empirical investigation of Maslow's theory. *Children and Youth Services Review, 34*, 1862–1867.
Northern Territory Environment Protection Authority. (2014). Waste and pollution. Retrieved August 5, 2014 from http://www.ntepa.nt.gov.au/.
NSW Government. (2010). NSW waste avoidance and resource recovery strategy. discussion draft: strategic directions and implementation plan 2011–2015. Retrieved August 23, 2014 from http://www.environment.nsw.gov.au/resources/warr/101029stratdirimpplan.pdf.
Queensland Government. (2014). Retrieved August 5, 2014 from http://www.derm.qld.gov.au.
Roads and Maritime Services. (2014). Roads and maritime. Retrieved August 5, 2014 from http://www.rms.nsw.gov.au/index.html.
Saeednia, Y., & Nor, M. M. (2013). Measuring hierarchy of basic needs among adults. *Procedia—Social and Behavioral Sciences, 82*, 417–420.
Smith, L., Alles, C., Lemay, K., Reddel, H., Saini, B., Bosnic-Anticevich, S., et al. (2013). The contribution of goal specificity to goal achievement in collaborative goal setting for the management of asthma. *Research in Administrative Pharmacy, 9*, 918–929.
Sustinability Victoria. (2014). About us. Retrieved August 5, 2014 from http://www.sustainability.vic.gov.au.
Tam, V. W. Y., & Tam, C. M. (2008). Waste reduction through incentives: A case study. *Building Research & Information, 36*, 37–43.
Taswaste. (2014). Rethink waste Tasmania. Retrieved August 7, 2014 from http://www.taswaste.com/.
The NSW Environment Protection Authority (EPA). (2007). NSW waste avoidance and resource recovery strategy 2007. Retrieved August 1, 2014 from http://www.epa.nsw.gov.au/warr/WARRStrategy2007.htm.

VicRoads. (2014). VicRoads contract documents. Retrieved August 5, 2014 from http://webapps.vicroads.vic.gov.au/VRNE/csdspeci.nsf.

Waste Authority. (2014). Strategic Partner—MBAWA. Retrieved September 6, 2014 from http://www.wasteauthority.wa.gov.au/programs/partners/strategic-partnerships/strategic-partner-mba.

Zaman, A. U., & Lehmann, S. (2011). Challenges and opportunities in transforming a city into a "zero waste city". *Challenges, 2*, 73–93.

Chapter 2
Scholars' Comments and Criticisms on Construction and Municipal Solid Waste Management in Hong Kong and Singapore from 2000 to Present: A Review of Digital Information Online

Rita Yi Man Li, Simon Fong and Hoi Man Chan

Abstract Waste is considered as an end of city product's life as well as a burden in modern society. The government of Hong Kong and Singapore are striving to improve construction and municipal solid waste management since 1960. At the same time, waste policies lead to social and economic controversies among citizen. How can we develop better sustainable waste management strategies? Will these methods lead to different level of environmental and economic problems? Whilst previous research in construction and municipal waste mainly sheds light on the methods of construction waste management on sites by interviews, surveys and case studies, few research has been conducted to study the digital information available in academic journal database and World Wide Web. This chapter tries to fill this gap by discussing (1) whether the government's waste management strategies are sustainable; (2) the efforts on construction and municipal solid waste made by the governments in Hong Kong and Singapore.

Keywords Municipal solid waste management · Hong Kong · Singapore · Landfill · Recycling · Sustainable development · Construction waste

R.Y.M. Li (✉)
Department of Economics and Finance / Sustainable Real Estate Research Center, Hong Kong Shue Yan University, Hong Kong, China
e-mail: ritarec1@yahoo.com.hk

R.Y.M. Li
School of Natural and Built Environments, University of South Australia, Adelaide, Australia

S. Fong
Faculty of Science and Technology, University of Macau, Macau, China

H.M. Chan
Department of Economics and Finance, Hong Kong Shue Yan University, Hong Kong, China

R.Y.M. Li, *Construction Safety and Waste Management*,
Risk Engineering, DOI 10.1007/978-3-319-12430-8_2

2.1 Introduction

The fast economic and population growth in Hong Kong and Singapore has led to a large amount of waste, especially construction and municipal solid waste (MSW). The increase in the amount of waste causes serious environmental burden such as natural habitat degradation, pollution of air, land, and water. Citizens, businesses, and the local government are now concern of the waste issues. In order to achieve a higher level of resource efficiency, strategic waste management on MSW and construction waste is indispensable (Tanaka 2007; Feng et al. 2009). In general, Lehmann et al. (2013) suggest that effective waste management strategies should include the following 5 major actions:

(1) Eliminate waste consignment to landfill;
(2) Advance the process of resource recovery and recycling;
(3) Adopt suitable technology is used to increase the resources recovery and reduce the amount of waste disposal to landfills;
(4) Improve waste sorting facilities' efficiency and effectiveness;
(5) Improve materials' recovery according to the Environment Protection Policy so as to turn the waste to resources.

Nevertheless, different waste strategies result in different level of impacts on economy and environment. That is why the government of Hong Kong and Singapore has been committed to solve the problems of construction and municipal solid waste since 1960. To manage the waste problems, up-to-date information is needed. Digital information plays an important role in receiving and sharing the knowledge of various construction waste management strategies. Nevertheless, previous research on construction and municipal waste mainly rest on the methods and difficulties encountered by stakeholders via interviews, case studies and surveys. Few have done research on the digital information available in our internet. This chapter aims to fill this gap of research. It studies the information in academic journal databases and World Wide Web.

2.2 Sustainable Development

The concept of sustainable development is gaining popularity in recent years and there are many different definitions from different school of thought. The idea of sustainable development is synonymous with sustainability. It is derived from an older forestry term "sustained yield", which is a translation of a German term in 1713 (Donovan 2009). The concept of sustainability balances between reproduction and resource consumption already existed in the 12th century. However, the most widely quoted definition of sustainable development came from the Brundtland report in 1987 "sustainable development is development that meets the needs of the

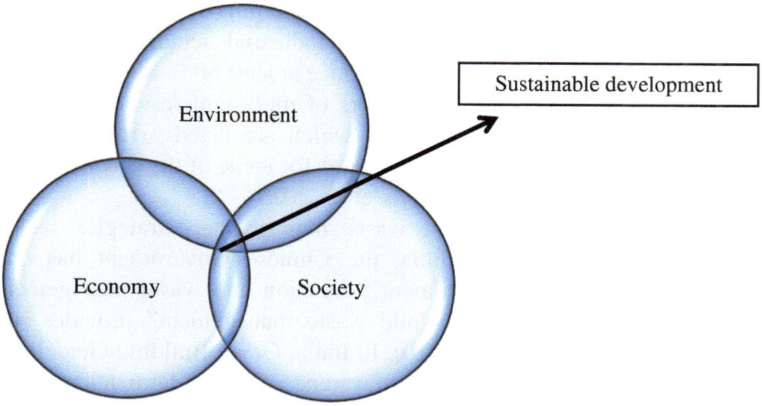

Fig. 2.1 Sustainable development: a co-development of economic, social and environment (Li 2011)

present without compromising the ability of future generations to meet their own needs" (World Commission on Environment and Development 1987).

To achieve the goal of sustainable development according to the World Commission on Environment and Development (1987), an organization/individual must fulfil the following three major criteria:

1. Efficient: meeting our own needs and aspirations without doing damage to the prospects of future generations;
2. Clean: finding ways to increase prosperity and improve the quality of life while reducing overall pollution and waste;
3. Green: reducing the environmental burden we put on our neighbors and helping to preserve common resources (Sustainable Development Unit 2010; Sustainable Singapore 2013).

Another major view of sustainable development suggests that it embraces the co-development of social, environment and economics (Li 2010, 2014; Li and Ah Pak 2010). As it is used by the Hong Kong and Singapore government, our chapter will also adopt this version to discuss the issue of sustainability from this perspective (Fig. 2.1).

2.3 A Global Perspective on Waste and Its Management

In Canada, Ontario Regulation 102/94 mandatorily requires business or building owners to develop and implement waste reduction plans, update the audits and plans annually and conduct waste audits. Later, Ontario Regulation 103/04 revises

Ontario Regulation 102/94 (Yeheyis et al. 2013) and impose the 3Rs Regulations to ensure that municipalities, commercial and industrial sector to develop waste disposal reduction strategies which aim to reduce at least 50 % of the waste material (Yeheyis et al. 2013). It requires the owners of multi-unit residential (apartment) buildings with six or more units, owners which are listed in this regulation to implement source of waste separation programs for some of the specific wastes and to ensure that waste is recycled (Yeheyis et al. 2013).

In BRICS, various construction waste management strategies have been implemented. For example, in China, the Chinese Government has enforced regulations with regards to environment protection and waste management. For example, "measures for municipal solid waste management" provides rules on waste disposal charges (Hao et al. 2010). In India, Green Building Guidelines such as LEED India has led to a growing awareness in standardization of design, over-ordering minimization, workers' environmental education, increase offsite construction materials' production, just in time deliveries, waste auditing (Arif et al. 2012). In Delhi, there is over 120,000 *Kabari* (informal waste collectors) on the streets who collect aluminum cans, paper, plastic, glass and sell them to mini-scrap dealers as secondary raw materials market. This informal industry processes 59 % of Delhi's waste and provides economic support to many families. In Curitiba, Brazil, the "Green Exchange Programme" offers slum dwellers fresh vegetables and free bus tickets when they collect garbage a to neighbourhood centres. Children are encouraged to exchange materials which can recycle toys (Lehmann 2011).

2.4 A Tale of Two Cities: Waste Management in Hong Kong and Singapore

This chapter studies the construction and municipal solid waste management in Singapore and Hong Kong and discusses possible solutions on how we can improve sustainable development. Although Singapore is an independent country and Hong Kong is only a small city in China, they share similar characteristics such as economic structure and situation, once ruled by the UK as British colonies and culture (both of them have a high proportion of Chinese).

Geographically speaking, Hong Kong and Singapore are small, densely populated coastal developing city-states located in Asia with limited natural resources. Their superb locations and excellent harbors caused the boom of trading in the early nineteenth century and population grew rapidly. According to the 2011 Population density report by the World Bank, both Hong Kong and Singapore rank the top-five of the most densely populated areas (World Bank 2014).

Likewise, they share similar economic development. For example, the 2014 annual report of economic freedom by Heritage Foundation, Hong Kong and Singapore are two out of the four places with the freest economy (The Heritage Foundation 2014). Besides, both of them share similar economic system and development. The World Bank shows that the 2012 GDP per capita at purchasing power parity of Hong Kong and Singapore stood at USD 61,803 and USD 51,945 which were ranked as the 5th and 8th among the world highest GDP per capita (World Bank 2014).

2.5 Waste Statistics in Hong Kong and Singapore

With rapid economic development, Hong Kong and Singapore are now facing the problem of increasing waste. In 2011, there was about 6.3 million tonnes of waste in Hong Kong. Each person generated around 463.5 kg of waste per year. The major source of waste in landfills is municipal solid waste (66 %), which included food waste, metals, plastic, etc. The second main component is construction waste that is shown in Fig. 2.2. Besides, the total waste generated has increased by almost 50 % from 4.26 million tonnes in 1991 to 6.3 million tonnes in 2011. At the same time, the construction waste has a remarkable reduction at landfills that dropped from 5.98 million tonnes in 1991 to 1.21 million tonnes in 2011, approximate 80 % decrease. However, the quantity of MSW disposed of at landfills has raised 21 % since 1991 (Hong Kong Environmental Protection Department 2011) (Fig. 2.3).

In Singapore, there was about 6.9 million tonnes of waste generated in 2009 and i.e. each person generated around 1,330 kg of waste in a year. The total waste generated increased by 56 % from 4.6 million tonnes in 2000 to 7.3 million tonnes in 2012 (Fig. 2.4). Majority is MSW, contributing 70 % of all kinds of waste (Fig. 2.5) (Singapore National Government Agency 2013). The proportion of different type of the waste in Hong Kong and Singapore are similar.

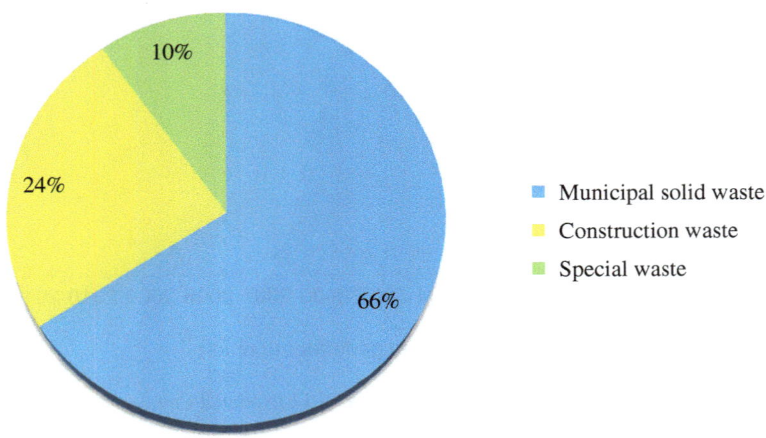

Fig. 2.2 Hong Kong disposal of solid waste at landfills in 2011 (Environmental Protection Department 2011)

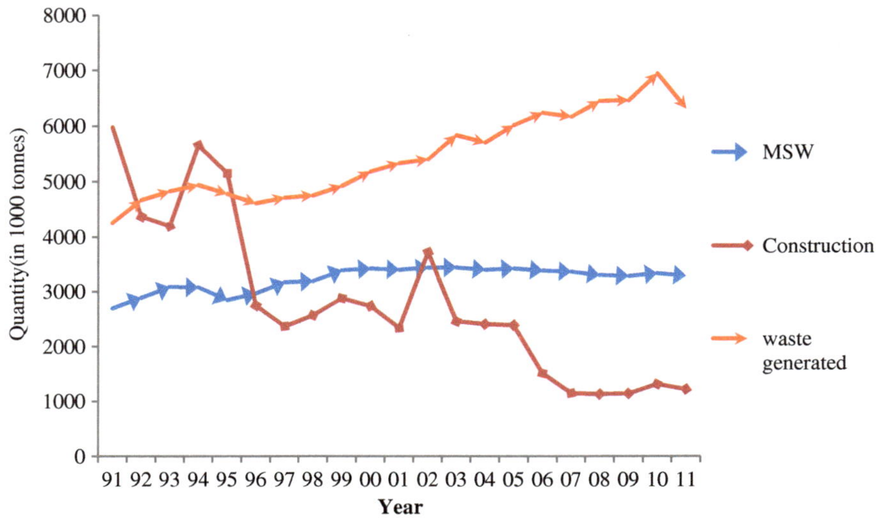

Fig. 2.3 Quantities of solid waste generated in 1991–2011 in Hong Kong (Environmental Protection Department 2011)

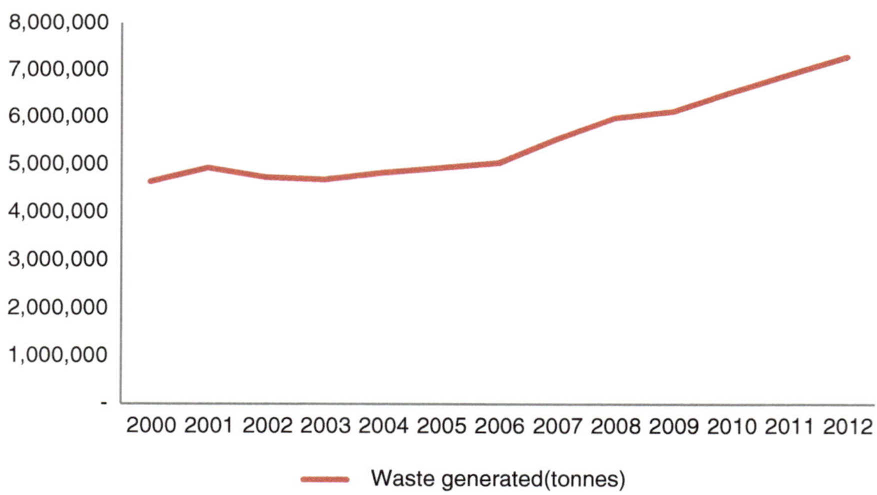

Fig. 2.4 Volume of waste in Singapore from 2000 to 2012 (National Environment Agency 2013)

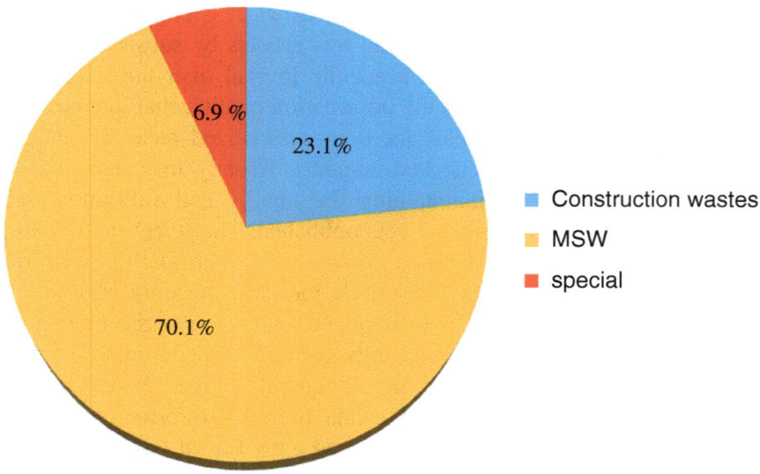

Fig. 2.5 Distribution of waste in Singapore in 2011 (National Environment Agency 2013)

2.6 Digital Information

Before the era of information digitization, a researcher who wanted to read an academic journal would have to travel down to a local library and access to the journal archive. The archive by the constraint of physical space available, maintains only a limited volumes of journals where some old issues may have been purged for making space for the new ones. The problems of latency in accessing paper journals in a physical premise, and incomplete information have been improved in the early 90s when a movement of digitizing journal information and founding online journals emerged. For instance, Mr. William Bowen, the president of Princeton College founded an online journal library called Journal Storage (JSTOR) that digitized academic journals from different disciplines into electronic copies, made available for online access. There are other popular and open-access digital journal archives such as, AGRIS, arXiv, CiteSeerX, ERIC, LexisNexis, NSDL.org, just to name a few.

To librarians, the digitization translates to savings on operation costs in managing hard-copies of the journals and freeing up precious physical storage space. To the users, it is the convenience in finding journal articles of many different publishers, from the online journal databases instead of physical journal shelves and rooms. There are some quality digital journal libraries providers, such as JSTOR that charges subscription fees for the license of use to colleagues, universities or commercial research institutes. The licensees are allowed to provide online access, usually via Intranet, to their students and staff in doing research usually at no charge. Some selected popular digital journal operators that charge subscription fees include but not limited to, ACM digital library, Bowker, CINAHL, Econ Lit, IEEE Xplore and POIESIS etc. for accessing full text articles by subscription.

The advantage of digitized online academic journal over the free but uncontrolled materials from the Internet is the information quality, that includes accuracy and authoritivity. Academic journals focus on specialized areas or categories of information, with meticulously checked contents. Usually in a reputable journal, peer review is conducted for maintaining the quality and substantial scientific credentials each paper carries before it gets published. The scientific journal papers usually contain credible data, described empirical research in full and formalized theories. The information from these scholarly journal papers are meant to be used by fellow researchers for on-going subsequent research. On the other hand, the information found on the Internet is usually unverified, and sometimes biased in views with either underlying political or commercial motives. Some are published for entertainment sake or with a propaganda purpose. Informative these online reports may be, yet mostly are gossips like and even full of advertisements.

In summary, researchers are offered the convenience of collecting relatively more reliable data from online digital journal archives than from websites on Internet. It takes only a computing device with Internet connection to do data collection rather than physically visiting a library. The other advantage lies on the ease and wide variety of available online databases for searching. The full-text article in each journal database can be effortlessly downloaded, stored on a mobile device and read at any time anywhere later as the research wishes. This can significantly reduce the hassles in academic information searches, leading to more focused research effort hence better results.

2.7 Research Method

In this chapter, we mainly discuss the most common and popular methods among these two places, i.e. landfill, recycle and incineration. Updated statistics about the operation of these three waste management methods is obtained from Environmental Protection Department (2011) in Hong Kong and National Environment Agency (2013) in Singapore.

Since this paper is a qualitative research on the comments and criticisms of the construction and municipal solid waste management. We adopt systematic content analysis as the research method. The scholarly literatures are searched and collected from the Google scholar, EBSCOhost research databases and ScienceDirect. The keywords use include 'Municipal solid waste management', 'Hong Kong', 'Singapore', 'landfill', 'recycling', 'sustainable waste management' from 2000 to present.

2.8 Similar Approaches of Construction Waste and MSW Management in Hong Kong and Singapore

Hong Kong and Singapore are densely populated cities in the world, waste has become one of the major problems that the government has to deal with. Both of them adopt similar method in handling waste, i.e. landfill and waste recycle.

2.8.1 Landfill

Landfill is the most popular method for waste disposal in the world, Hong Kong and Singapore are of no exception. In 1960, Hong Kong established the first landfill site in Kwai Tsing and the number of landfill increased to 16 by 1995. Nowadays, most of the landfills in Hong Kong were already closed and changed in use of land, only three landfill sites are still operating in the New Territories (Environmental Protection Department 2011). More than 90 % MSW and construction wastes are now disposed in three large landfills by waste collection vehicles every day.

Same as Hong Kong, there is an offshore landfill in the southern part of Singapore. The Semakau Landfill receives incineration ash which is delivered by tugboats at night. It also provides educational and interesting outings for local citizens, students and foreign visitors (National Environment Agency 2013). Even though Pulau Semakau operated a landfill, the western coast has become a scenic and green natural place with fresh air and provides a bub for rich biodiversity.

2.8.1.1 Scholars' Comments and Criticisms on Landfilling

Although landfill is the most widely use method to dispose MSW and construction waste, it is always considered as a costly, unsustainable way to our environment and economy. It is also not welcomed by the environmental groups. The European Union waste strategy consider landfill as the last resort of waste management (Depountis et al 2009). It is not environmentally sustainable owing to the potential hazard of organic materials (Bjarnadóttir et al. 2002). The landfill also gives rise to environment problems such as noise pollutions, emissions of dust and gases to atmosphere, contaminated surface water to watercourses (Depountis et al. 2009). However, the Semakau Landfill is operated in a more environmental friendly that very minimal gases existed in the landfill as organic wastes are not sent to landfill sites and there is a maximum 10 % of the total solid waste generated going to landfill under the National Environment Agency (NEA) scheme (Tan and Khoo 2006).

In Hong Kong, as the three landfills in Hong Kong has been used for about 20 years, the Environmental Protection Department (2011) estimated that all the existing landfills will be full by 2017 (Chung and Poon 1998). Landfill crisis has become a major challenge in Hong Kong owing to the limited land resources while

Table 2.1 Summary of scholars' comments on landfilling

Author (year)	Scholar comments	Place(s)
Fischer (1999)	Landfill is considered as the last resort in the waste management	Hong Kong and Singapore
Bjarnadottir et al. (2002)	It is not considered as environmentally sustainable owing to the potential release of hazard organic materials	Hong Kong and Singapore
Depountis et al. (2009)	Landfill would arise environment problems like emissions to atmosphere and water including noise, dust, gases, contaminated surface water runoff to watercourses	Hong Kong and Singapore
Tan and Khoo (2006)	The Semakau landfill is operated more environmental friendly with very minimal gases released at the landfill	Singapore
Tan and Khoo (2006)	Organic wastes should be disposed to the landfills	Singapore
Chung and Poon (1998)	Hong Kong is now facing an urgent problem of soon-to-be full landfill sites	Hong Kong
Yau (2010)	Landfill crisis has been seen as a major challenge in Hong Kong owing to the limited land resources. Creating new ones or extending the existing landfill sites is extremely costly	Hong Kong

opening new ones or extending the existing landfill sites is extremely costly (Yau 2010) (Table 2.1).

2.8.2 *Recycling*

In view of the urgent landfill crisis in Hong Kong and air pollution released from incineration plant in Singapore, recycling waste has been taken as an intermediary measure which creates less environmental damage and achieve environmental sustainability (Yau 2010). In Singapore, to support the construction waste recycle, the NEA provides lots of recycling facilities which turn the waste into non-structural concrete products for building construction (Zhang et al. 2009). According to the Singapore's NEA (2013), the recycle rate of construction waste increased from 85 % in 2001 to 99 % in 2012, reaching the target recycling rate set in Singapore Green Plan 2012 successfully. Besides, the recycling rate of MSW in 2012 is 58 %.

Owing to the rapid depletion in landfill sites, the Hong Kong government also set a number of Waste Reduction Task Forces to promote the idea of recycling in different sectors since 1997. At that moment, the recycle rate of domestic waste was around 8 % of the total waste disposed (Chan 1998). Hence, the government launched a 10 years implementation program in the Waste Reduction Framework Plan in 1998 (Environmental Protection Department 2011). The program includes provision of waste separation bins in public areas and encourages the developers to provide space for refuse storage and waste recovery (Environmental Protection Department 2011).

2.8.2.1 Scholars' Comments and Criticisms on Recycling

It is often of the view that recycle is a good way to achieve a sustainable development. It enhances the level of resource efficiency (Tanaka 2007), save energy, prolongs lifespans of landfills and reduces emissions to air, water from incinerators and landfills (Seik 1997; Ekins et al. 2003). Nevertheless, our stakeholders are fully aware of various economic concerns (Schultz et al. 1995). Yau (2010) comments that recycling can be a heavy long-term financial burden to the HKSAR Government if it offers monetary incentives to increase domestic waste recycle. Thus, the HKSAR Government should consider incorporating different economic incentives to encourage waste recycle such as setting up a reward scheme so that citizens can exchange recyclable materials for free public services. Besides, Hong Kong's recycle market is limited to recycling materials with a high scrap value and lacks of recycle facilities which may create hazardous substances (Tam and Tam 2006).

In Singapore, Zhang et al. (2009) believed that the Singapore Government endeavors to reduce waste reduction from disposing in the landfill by recycling. Hence, the lifespan of the Semakau Landfill increases from 20–30 years to 35–40 years. Even though Singapore successfully recycles waste, costs of recycling are high because of high waste collection and handling costs (Bai and Sutanto 2002) (Table 2.2).

Table 2.2 Summary of scholars' comments on waste recycle

Author (year)	Scholar comments	Place(s)
Tanaka (2007)	Waste recycle can achieve a high level of resource efficiency	Singapore and Hong Kong
Seik (1997), Ekins et al. (2003)	Recycle saves energy, prolongs life spans of landfills and reduces emissions to air, water from incinerators and landfills	Singapore and Hong Kong
Yau (2010)	Waste recycle would become a heavy long-term financial burden to the Hong Kong Government if it offers inducements continuously to boost domestic waste recycling. The Hong Kong Government should consider incorporating different economic incentives in its policies to encourage waste recycling like set up a reward scheme that citizens can exchange recyclable materials for free public services	Hong Kong
Tam and Tam (2006)	Hong Kong's recycling market is limited due to a high scrap value od the recycled materials and lacks of recycling facilities for waste which may create hazardous substances	Hong Kong
Zhang et al. (2009)	Singapore Government have helped tremendously to reduce waste from disposing in the landfill, as a result, the life span of the Semakau landfill increases from 20–30 years to 35–40 years	Singapore
Bai and Sutanto (2002)	The cost of recycling has been proven to be higher than original anticipation because of high waste collection and handling cost	Singapore

2.9 Different Approaches of Construction and MSW Management in Hong Kong and Singapore

2.9.1 Incineration

Solid waste incineration has been placed in a top priority over the past 30 years in Singapore. Incineration is a waste treatment process that converts the waste into ash, heat and flue gas and finally the ash would be disposed at the landfill. The incineration plants are known as waste-to-energy (WTE) plants because the steam generated during the combustion process would be used to produce electricity. The recovered energy is mainly used to operate the waste-to-energy plant, and the extra ones are then sold to the electricity institutions. As a result, the revenue from the sale of energy becomes a major income to the plants (Bai and Sutanto 2002). According to NEA (2013), there are five incineration plants in Singapore currently. They are Ulu Pandan (1,600 tons/day), Tuas (2,000 tons/day), Senoko (2,400 tons/day) and Tuas South (3,000 tons/day). At present, the solid waste disposed in Singapore is about 7 million tonnes and 37.6 % of the waste is going to waste-to-energy plants for energy recovery and incineration (National Environment Agency 2013).

2.9.1.1 Scholars' Comments and Criticisms on Incineration

Rapid industrialization and economic development causes tremendous increase of solid waste incineration in Singapore, even though it is much more expensive method than landfill (Bai and Sutanto 2002). Nevertheless, incineration has long been considered as non-sustainable waste management. According to Hellweg et al. (2001), wastes incineration generates nitrogen oxides, which could potentially contribute to environmental problems like acidification and ecotoxicity. Also, the ashes generated from incineration always show substantial levels of toxic substances (Qian et al. 2006). Therefore, Khoo et al. (2012) suggested NEA to find alternative uses for incineration ash instead of disposing at landfills. Therefore, Singapore tested the bottom ash to see whether it could be used as road pavement in order to reduce the amount of material sent to landfills (Tan and Khoo 2006). Although the incineration plant can produce electricity during waste burning, Tan and Khoo (2006) commented that the energy gained is outweighed by the environment damage caused by the incinerators (Table 2.3).

Table 2.3 Summary of scholars' comments on incineration

Author (year)	Scholar comments	Place(s)
Bai and Sutanto (2002)	Incineration is much more expensive method than landfilling	Singapore
Hellweg et al. (2001)	Incineration of wastes would generate emissions such as nitrogen oxides, which could potentially contribute to environmental problems like acidification and ecotoxicity	Singapore
Qian et al. (2006)	The ashes generated from incineration always show substantial levels of toxic substances	Singapore
Khoo et al. (2012)	The government should find alternative uses of incineration ash instead of disposing at landfills	Singapore
Tan and Khoo (2006)	Singapore tested the bottom ash to see whether it could be used as road pavement in order to reduce the amount of material sent to landfills	Singapore
Tan and Khoo (2006)	The energy gained is outweighed by the environment damage caused by the incinerators	Singapore

2.10 Summary

Sustainable construction and solid waste management requires full consideration in economic, social and environmental aspects. Even though landfill and incineration have long been considered as an unsustainable waste management, Singapore shows how waste disposal management can be operated in a more sustainable way. The Semakau landfill provides a good example to the world and change people's views on landfills. It not only provides space to store wastes, but also a comfortable place with educational value, fresh air and rich biodiversity to its citizens and visitors. At the same time, Singapore puts incineration in a top priority as an waste management strategies. During the process of waste combustion, the steam generated would be used to produce electricity that generates economic benefits.

On the other hand, Hong Kong still relies on the traditional, inefficient landfill as a final destination for construction and municipal solid waste. It is now facing a serious problem of landfill sites shortage. The expansion of landfill sites has led to a essential controversial discussion in the city and it is in no doubt a hard nut to crack issue for the Hong Kong government.

In view of the alarming landfill crisis and the environment damage from incineration, the Hong Kong and Singapore government promote waste recycle. Among all the suggested solutions for mountains of waste, recycle has been considered as one of the most viable solutions as evaluated by academic scholars as it achieves a high level of resource efficiency (Tanaka 2007) and reduces emissions of air and water pollutants from incinerators and landfills (Seik 1997; Ekins et al. 2003). However, if the government does not consider any other different economic incentives, recycle waste can impose heavy long-term financial burden. Hence, it is not a sustainable solution.

To achieve sustainable development, construction and municipal solid waste disposal may impose a major obstacle in city development. Therefore, in order to strike a balance between economic growth, social well-being and environmental protection, it is necessary and essential for the Singapore and Hong Kong government to promote recycle in an effective way and consider more sustainable waste management strategies.

References

Arif, M., Bendi, D., Toma-Sabbagh, T., & Sutrisna, M. (2012). Construction waste management in India: An exploratory study. *Construction Innovation, 12*, 133–155.

Bai, R., & Sutanto, M. (2002). The practice and challenges of solid waste management in Singapore. *Waste Management, 22*, 557–567.

Bjarnadóttir, H. J., Fri Riksson, G. B., Johnsen, T., & Sletsen, H. (2002). *Guidelines for the use of LCA in the waste management sector*. http://www.norsas.no/content/download/219/969/file/Guidlines%20for%20the%20use%20of%. Accessed 17 May 2014

Chan, K. (1998). Mass communication and pro-environmental behaviour: Waste recycling in Hong Kong. *Journal of Environmental Management, 52*, 317–325.

Chung, S. S., & Poon, C. S. (1998). A comparison of waste management in Guangzhou and Hong Kong resources. *Conservation and Recycling, 22*, 203–216.

Depountis, N., Koukis, G., & Sabatakakis, N. (2009). Environmental problems associated with the development and operation of a lined and unlined landfill site: A case study demonstrating two landfill sites in Patra, Greece. *Environmental Geology, 56*, 1251–1258.

Donovan, F. (2009). *Our uncertain future: Can good planning create sustainable communities?* US: ProQuest UMI Dissertation Publishing.

Ekins, P., Simon, S., Deutsch, L., Folke, C., & de Groot, R. (2003). A framework for the practical application of the concepts of critical natural capital and strong sustainability. *Ecological Economics, 44*, 165–185.

Environmental Protection Department. (2011). *Environmental protection department*. Retrieved December 13, 2013, from http://www.epd.gov.hk/epd/eindex.html.

Feng, H., Kollikkathara, N., & Stern, E. (2009). A preview of waste management evolution: Special emphasis on USA. *Waste Management, 29*, 974–985.

Hao, J. L. J., Tam, V. W. Y., Yuan, H. P., Wang, J. Y., & Li, J. R. (2010). Dynamic modeling of construction and demolition waste management processes: An empirical study in Shenzhen, China. *Engineering, Construction and Architectural Management, 17*, 476–492.

Hellweg, S., Hofstetter, T. B., & Hungerbuhler, K. (2001). Modeling waste incineration for life-cycle inventory analysis in Switzerland. *Environmental Modelling and Assessment, 6*, 219–235.

Khoo, H., Tan, L. Z., & Tan, B. H. (2012). Projecting the environmental profile of Singapore's landfill activities: Comparisons of present and future scenarios based on LCA. *Waste Management, 32*, 890–900.

Lehmann, S. (2011). Optimizing urban material flows and waste streams in urban development through principles of zero waste and sustainable consumption. *Sustainability, 3*, 155–183.

Lehmann, S., Zaman, A. U., Devlin, J., & Holyoak, N. (2013). Supporting urban planning of low-carbon precincts: Integrated demand forecasting. *Sustainability, 5*, 5289–5318.

Li, R. Y. M. (2010). Traditional Chinese mode of thinking, sustainable development and property management. *Real Estate and Development Economics Research Journal, 2*, 70–86.

Li, R. Y. M. (2011). *Building our sustainable cities*. Illinois: Common Ground Publishing.

Li, R. Y. M. (2014). *Law, economics and finance of the real estate market—a perspective of Hong Kong and Singapore*. Germany: Springer.

Li, R. Y. M., & Ah Pak, D. H. (2010). Resistance and motivation to share sustainable development knowledge by Web 2.0. *Journal of Information and Knowledge Management, 9*, 251–262.

National Environment Agency. (2013). *Waste management*. Retrieved October 10, 2013, from http://app2.nea.gov.sg.

Qian, G., Cao, Y., Chui, P., & Tay, J. (2006). Utilization of MSWI fly ash for stabilization/solidification of industrial waste sludge. *Journal of Hazardous Materials, 129*, 274–281.

Schultz, P. W., Oskamp, S., & Mainieri, T. (1995). Who recycles and when? A review of personal and situational factors. *Journal of Environmental Psychology, 15*, 105–121.

Seik, F. T. (1997). Recycling of domestic waste: Early experiences in Singapore. *Habitat International, 21*, 277–289.

Singapore National Government Agency (2013). Safeguard, Nurture, Cherish. http://app2.nea.gov.sg/. Accessed 28 June 2014

Sustainable Development Unit (2010). What is Sustainable Development. http://www.susdev.gov.hk/html/en/sd/index.htm. Accessed 13 December 2013

Sustainable Singapore (2013). Sustainable Singapore A Lively and Liveable City. http://app.mewr.gov.sg/web/contents/ContentsSSS.aspx?contid=1307. Accessed 28 June 2014

Tam, W. Y., & Tam, C. M. (2006). Evaluations of existing waste recycling methods: A Hong Kong study. *Building and Environment, 41*, 1649–1660.

Tan, R. H., & Khoo, H. H. (2006). Impact assessment of waste management options in Singapore. *Journal of the Air and Waste Management Association (Air and Waste Management Association), 56*, 244–254.

Tanaka, M. (2007). Waste management for a sustainable society. *Journal of Material Cycles and Waste Management, 9*, 2–6.

The Heritage Foundation. (2014). *2014 index of economic freedom*. Retrieved May 17, 2014, from http://www.heritage.org/index/press-release-asia-pacific.

World Bank. (2014). *Data*. Retrieved May 17, 2014, from http://www.worldbank.org/.

World Commission on Environment and Development. (1987). *Our common future*. Retrieved May 17, 2014, from http://www.un-documents.net/ocf-02.htm.

Yau, Y. (2010). Domestic waste recycling, collective action and economic incentive: The case in Hong Kong. *Waste Management, 30*, 2440–2447.

Yeheyis, M., Hewage, K., Alam, M. S., Eskicioglu, C., & Sadiq, R. (2013). An overview of construction and demolition waste management in Canada: A lifecycle analysis approach to sustainability. *Clean Technologies and Environmental Policy, 15*, 81–91.

Zhang, D., Tan, S. K., & Gersberg, R. M. (2009). A comparison of municipal solid waste management in Berlin and Singapore. *Waste Management, 30*, 921–933.

Chapter 3
A Content Analysis on Static Digital Information About the Causes of Construction Accidents

Rita Yi Man Li, Sun Wah Poon and Daniel Chi Wing Ho

Abstract In many places, construction accident rates are high. While construction companies incur a huge sum of money for compensation in accidents, the injured workers suffer from loss of income and even capacity to work. As implementation of construction safety measures and relevant regulations require the knowledge of construction accident causes, this research study aims at reviewing the digital information available in academic journal database about construction accidents with the help of information technology tools. The research results show that (1) many of the journal articles shed light on developed countries, with the majority come from Europe and the US. (2) Heaps of research analyses the statistics from official database or report as their method of research studies, followed by questionnaires/survey or interview. Triangulation, i.e. more than one research method, is seldom used in these journal articles. (3) Science Direct and ASCE library contain more journal articles which are related to causes of construction accidents than the other databases. (4) Lots of them concede that accidents is more likely to happen because of the immediate causes of accidents, such as fall, slip and struck by objects, and the workers' characteristics.

Keywords Digital information · Construction accident causes · Academic journal database · Content analysis · Construction safety

---

R.Y.M. Li (✉)
Department of Economics and Finance / Sustainable Real Estate Research Center, Hong Kong Shue Yan University, Hong Kong, China
e-mail: ritarec1@yahoo.com.hk

R.Y.M. Li
School of Natural and Built Environments, University of South Australia, Adelaide, Australia

S.W. Poon · D.C.W. Ho
Department of Real Estate and Construction, The University of Hong Kong, Hong Kong, China

R.Y.M. Li, *Construction Safety and Waste Management*,
Risk Engineering, DOI 10.1007/978-3-319-12430-8_3

3.1 Introduction

The evolving industrial-based society into the knowledge-based society has made a huge change in the social environment. Researchers have witnessed an increasing value of knowledge and a rise in volume of information (Al-Rawajfih et al. 2010). The popularity of World Wide Web has altered the way we share and receive information. We are all moving from the mountains of paper to the digital World. Modern companies, for example, are taking advantage of our digital World by collecting, using, utilizing and processing corporate information online (Al-Htaybat 2011). Digital technology which combines the swift of communications worldwide has markedly increased the possibilities for global exchange and production (Das 2010). Digital information management system has played an important role in spatial information storage (Kerdvibulvech and Yin 2014; Seo and Park 2014), modern health care management system (Sui et al. 2013). It also opens up the new Islamic world's entrepreneurial opportunities (Frazier and Ewing 2009). It also used as a means of communication among students, lecturers, faculties, alumni and guests (Jabar et al. 2014). It is, however, estimated that a quarter of all printed classified advertisement will be lost to digital media in the coming decade (Lambert 2007). Young people today may view digital media on Web 2.0, such as Youtube, blogs, MySpace as a primary vehicle in obtaining information and print media as the add-on.

With the popularity of online electronic information, we can use any of the domestic network or even international information resources from another library overseas. It means that a library is no longer restricted on the museum collection of documents but enlarges to national or global information, volume of hard copies collection is no longer the major indicator to measure the value of libraries. Rather, the electronic database may hold a lot more information. A static repository of printed material has changed to an open, orderly, multi-functional and flexible information service system (Zhang 2011). Likewise, modern construction safety researchers may spend more time on searching for digital information than traditional printed materials. Having said that, we may wish to know what exactly can be found online. For example, can we find the causes of construction accidents throughout the whole World or the materials mainly focus on some particular areas only? What kind of the research methods have been used by the researchers (as that definitely affect the research quality and validity of the papers)? And more importantly, what is the current state-of-art with regards to the causes of accidents? In this chapter, we wish to shed light on digital information we obtain from academic journal databases. It firstly discusses (1) the reasons why we have to know more about the causes of construction accidents under the lens of economics, (2) a technical perspective on digital information and (3) the distribution of journal articles which cover the causes of accidents (geographical locations, database and research methods) and (4) the causes of accidents under the lens of academic scholars.

3.2 The Costs of Construction Accidents

Construction safety has always been the concern of many practitioners and academics. High fatalities and accident rates impose negative impact on construction industry's image, deterring talented workers to enter the industry. It also leads to pain, suffering and loss of the injuries and their families (Li and Poon 2007, 2009, 2013). The cost of construction accidents reached 10 % of the total labour costs (Laufer 1987). Saram and Tang (2005) showed that the pain and suffering costs of the victims of construction accidents in Hong Kong amounted to the payment of $219 million in 1999. From 2004 to 2008, the highest sum of one accident compensation case alone in High Court reached HK$13.8 million (Li and Poon 2009).

3.3 Digital Information: A Technical Perspective

Digital information is firstly transformed to binary image which is then encrypted using decimal expansion of irrational numbers and multiple Fourier transforms. The decimal expansion of irrational number is then used to generate random phase masks which can be used to scramble the binary image into random noise. The digital information displays as binary data stream and can be processed by computers, followed by transmission in the Internet (Lu and Cao 2013). In general, there are two types of digital information: static and interactive digital information. Static digital information includes digitized e-books. Interactive digital information refers to digital information products with an interactive user interface which resembles a typical website are termed interactive e-books. Interactive digital information refers to digital information products with an interactive user interface which resembles a typical website (Kirk et al. 2012).

3.4 Research Method: Content Analysis

Content analysis is used for the research. It has been applied in analyzing government documents (Awang et al. 2013), and qualitative interviews (Ahmad and Seet 2009; Er 2011). This research method did not appear until 1941 when printed mass media increased substantially in the US at the beginning of the 20th century. It resulted in many efforts to create objective, simple and scientific methods of analyzing news articles. Subsequently, the methods were taken up by other fields such as history, psychology, anthropology and linguistics. In content analysis, similar information is grouped into similar categories so as to create systematic and objective criteria for transforming written text which can be analyzed for symbolic content of communication (Li 2013).

3.5 Data Description

The journal articles are found in five academic journal databases: Wiley Blackwell, Taylor and Francis, Science Direct, Emerald, ASCE library in Run Run Shaw Library, and City University of Hong Kong. We use the keyword construction accident causes to search for the relevant journal articles. A total of 71 freely accessed and relevant articles are included in our present study. More than half of them come from Science Direct (38), followed by ASCE (19) and Wiley Blackwell (8) (Fig. 3.1). There are 24 articles which study the causes of accidents in the US, followed by the 17 articles in European Union (7 from UK, 6 from Spain, 2 for Finland, 1 for Denmark, 1 for Italy), 8 articles from Taiwan (Table 3.1). The results show that digital information held by these electronic databases (with free access right in this university) focus on developed countries in two continents, Europe and North America. Even though the internet is open to everybody with access to the World Wide Web, we may not know the causes of construction accidents in the developing World well. On the other hand, however, it simply reflects the truth that there are substantially more academic scholars in developed countries which study the causes of accidents than the developing countries on this issue. In terms of research method, majority of them analyze the official data from government statistics or reports. 16 of them use interviews and 11 use questionnaires/survey to figure out the causes of construction accidents. The number of articles adopt research triangulation in spite of its capability in overcoming the shortcomings of any of the abovementioned research methods (Li and Poon 2013) (Fig. 3.2).

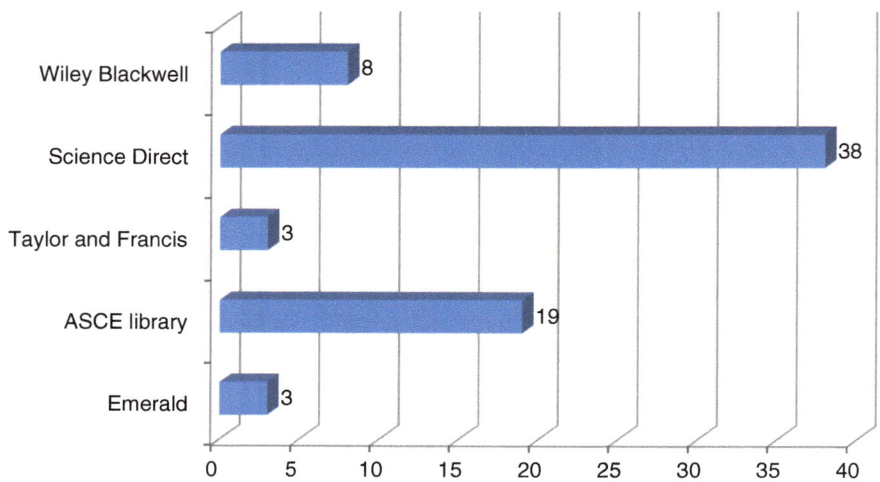

Fig. 3.1 Academic databases which have been covered in the research study

Table 3.1 Countries which are covered in the articles

Country	Number of articles
Australia	1
Brazil, Greece, Turkey, Portugal	1
Canada	2
China	3
China and Australia	1
Denmark	1
Finland	2
Hong Kong	2
Israel	1
Italy	1
Korea	1
Kuwait	1
Netherlands	1
Not specific	4
Singapore	1
Spain	6
Sweden and Denmark	1
Taiwan	8
Uganda	1
UK	7
United Arab Emirates	1
US	24

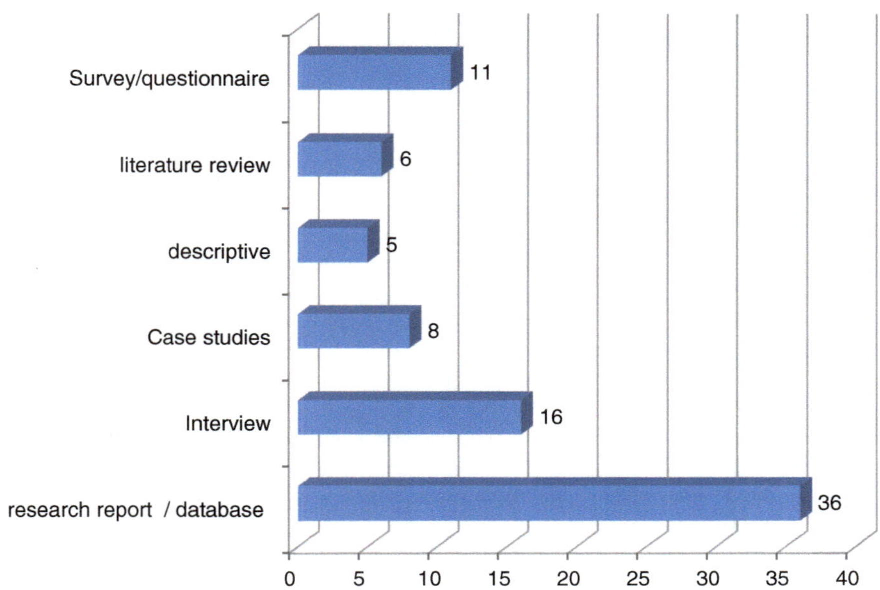

Fig. 3.2 Research methods used by the journal articles' writers

3.6 Causes of Construction Accident

The causes of accidents are subdivided into six categories: workers' characteristics/ background, organization, site condition, time of accidents, immediate direct causes which lead to accidents, and economic factors. Detail description of each of the category can be found in Table 3.2. Among all these factors, 33 of the research papers are concerned with the study of the immediate causes which lead to accidents such as fall from height, slip and struck by objects. Twenty nine articles shed light on the relationship between construction workers' background/characteristics and the chances of accidents. Twenty five articles reveal that site conditions such as lighting, building design and structural failure are critical factors which lead to accidents on sites (Fig. 3.3). In sharp contrast, however, there is only one article which links the economic factors to injuries on site that accidents most common in private projects with project contract amounts of less than NTD 5 million (Cheng et al. 2012).

3.6.1 Workers' Characteristics

There are heaps of research which show that workers' characteristics, attitudes, occupations etc. are the major indirect factors which affect the chances of accidents. For example, Haslam et al. (2005) find that 70 % of accidents are caused by workers or their work team. Cheng et al. (2010) find that foreign workers are more accident prone. Pratt et al. (1997)'s research results suggest that workers of repair, craft and precision production; material and transportation moving; equipment cleaners, handlers helpers and laborer have the highest rate and frequency of

Table 3.2 Categories of causes of construction accidents

Category	Factors which are included in each of the category
Workers' characteristics/ background	Workers' background information, such as age, sex, experience, human error, migrant, lack of safety training/education, migrant, poor relationship with others
Organization	Companies background and methods of grouping the workers, e.g. large company, subcontracting, preplanning
Site condition	The construction sites' characteristics, e.g. building design, structural failure, lack of protective measures, poor housekeeping
Time of accidents	In the morning, afternoon, evening/Monday to Saturday/ month when accidents happen
Immediate direct causes which lead to accidents	Fall from height, slip, struck by objects etc.
Economic factors	Piece rate method of payment, low accident compensation costs, high transaction costs, developing countries

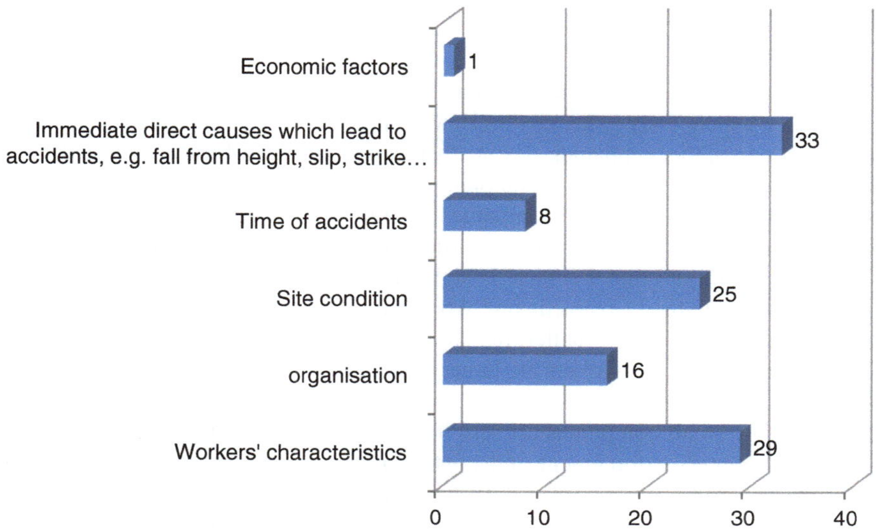

Fig. 3.3 Categories of the causes of construction accidents

fatalities. The number of accidents among workers aged between 41 and 60 is higher than those between 21 and 40 during Autumn and Summer (Liao 2012). Lin et al. (2011) notes that fatality rate is the highest among workers aged 34–44. Occupation also has implications on workers' safety risks. Arboleda and Abraham (2004) records that more than half of them aged between 19 and 40. Apart from age, other researchers focus on the experience of the workers. For example, Arquillos (2012) find that experienced workers do not have the lowest accident fatality rates. Memarian and Mitropoulos (2013b) points out that many of the masons' incidents happen during rebar and block laying activities. Block laying, Material handling, scaffold erection or dismantling often lead to high severity and frequency of incidents. Laborers in a masonry crew are at significantly higher risk of injury than the other positions such as foremen, operators and masons.

Workers' safety behavior (Hu et al. 2011), unsafe acts (Hale et al. 2012) and attitudes (Hu et al. 2011) affect the likelihood of accidents on sites. It is observed that repair, maintenance, alteration and addition works (RMAA) workers often underestimate potential safety risks when they work for small tasks for a short period of time (Hon et al. 2010). In China, the major safety risks came from low or no safety education paramount (Zou and Zhang 2009). Chan (2011) find that workers come across accidents because of fatigue.

3.6.2 Organization

Organization, such as size of the companies, organization policy, is another highly cited factor which leads to accidents (Hale et al. 2012; Langford et al. 2000). With regards to size of the companies, despite there are mountains of research which study the relationships between size of companies and accident rates, there is no consensus on whether small or large size companies have higher accident rates. Conventional research suggests that smaller size construction companies have higher injury rates. Cameron et al. (2008) find that there is higher percentage of fatality come from small contractors in Scotland. Arboleda and Abraham (2004)'s research shows that when the construction project is smaller, the chance of a trench cave-in is higher. Glazner et al. (1999) find that workers in large companies rarely report safety practices occurred all of the time, however, all of them report that some safety practices always occur. Small companies are less likely to indicate that these actions always occur. Apparently, enforcement and control affect employers' behaviors (Nelson et al. 1997). Nelson et al. (1997)'s data, however, find that a large company may not be safer than a small company in fatal accidents. OSHA devotes a substantial percentage of its resources to enforcement of very large contractors. It is also observed that enforcement activity has more substantial effects on contractor level, inspection effects seem to ripple across the sites controlled by contractors. Nevertheless, the positive marginal impact of repeat inspections usually happen on the initial compliance among large firms only if we perceive the issue from OSHA activity over a period of 15 years (Weil 2001). Moreover, it is found that the rate for workers' compensation claims for fall injuries drops after construction employers are cited for violating the fall protection standard (Nelson et al. 1997).

Langford et al. (2000) suggest that supervision, organization policy, and equipment management, risk taking and management behaviors affect the construction safety on sites. Hinze et al. (1998) show that 14.8 % of the respondents agree that site safety management affect the level of safety on sites. Dester and Blockley (1995) suggest that safety culture has not been developed is one of the causes of accidents. Rowlinson et al. (2003) find that senior management has not paid attention to the role of foreman on construction safety and poor site safety record has to be improved. Other organization problems which lead to construction accidents on sites include problems with suitability and condition of materials, shortcomings with equipment (including PPE) and deficiencies with risk management (Haslam et al. 2005). A research of 500 accident records provided by the U.K. Health and Safety Executive (HSE) shows that construction accidents involve inappropriate construction planning (28.8 %) and construction control (16.6 %) and operative action (29.9 %) (Suraji et al. 2001). In another research study, an analysis of 296 fatality reports on trenching operations from the Occupational Safety and Health Administration from 1997 to 2001 find that many of the accidents occur as sequence of procedure safe equipment or unsafe means of construction are not provided (Arboleda and Abraham 2004).

3.6.3 *Site Conditions*

Many of the construction sites are having poor site conditions problems (Hinze and Teizer 2011; Li and Bai 2008) and environmental issues (Hale et al. 2012). Hu et al. (2011) reveals that many of the accidents happen because of construction structure and facilities' problems on sites. Ale et al. (2008) suggest that the delivery system and barrier failure cause accidents. Li and Bai (2008) find that fatal accidents happen because of complicated road geometries and unfavorable light. Hu et al. (2011) concede that poor working surfaces and platforms cause accidents. Hinze and Teizer (2011) suggest that blind spots, lighting conditions and obstructions on sites lead to accidents. Dewlaney et al. (2012) indicate that design elements, methods and means of LEED building construction increase the exposure to high risk environments or chances of injuries. Suraji et al. (2001), however, has evidence that inappropriate site condition only account for 6 % of the accidents on sites.

3.6.4 *Time of Accidents*

Perez-Alonso et al. (2012)'s research show that Thursday has the greatest incidence of accidents (20.55 %), followed by Monday (20 %) whilst the period with the greatest number of accidents occurred in the first 4 h of the workday. Almost half of the accidents occurred close to the end of the week, i.e. Thursday and Friday. High rates of serious accidents occurs from 19:00 to 23:00 (Lópeza et al. 2011).

3.6.5 *Immediate Direct Causes Which Lead to Accidents*

3.6.5.1 Fall from Height, Slips and Collapse

Many of the previous research suggests that fall from height is one of the major direct causes of accidents (Huang and Hinze 2003; Zhou et al. 2012; Pinto 2014; Hinze et al. 1998; Wu et al. 2010; Irumba 2014). Cheng et al. (2012)'s research shows that fall or tumble was the most common cause of injury. It accounts for 47 % among all accident types. Many of the accidents happen due to fall from elevation, platform, ladder, roof and piled matter, fall on stairs and fall into openings (Nelson et al. 1997). Lin et al. (2011) find that approximately 30 % of fatal events are attributed to falls from stage or scaffold. Adam et al. (2009)'s survey find that 83 % of the interviewees state that many construction sites fail to provide protection nets underneath formwork against the consequence of a fall from a height. Leu and Chang (2013) studies the direct and indirect causes of falls. It is found that the most significant indirect causes of falls include improper use of safety equipment, improper usage of personal protective equipment, a lack of effective safety facilities, ineffective safety equipment, unqualified safety nets, and the non-existence of safety device. On the other hand, the most significant direct cause of falls is hoisting steel beams but lack of safety facilities. When hoisting steel beams, the major causes of falls include a lack of ideal fixed points for lifelines and safety belts, limited construction pedal boards and the inability to erect a safety net (Leu and Chang 2013). Weil (2001) records that while 28 % of all fatalities arose from falls on sites, 42 % of all the OSHA standards cited pertained to fall protection or falls. Previous research also shows that some seasons are more likely to have fall accidents. During winter, Liao (2012) suggests that the higher risk of fall-related incidents due to unpredictable wet weather conditions has to be managed with care. Weil (2001)'s research reveals that the percentage of standards cited which is related to falls (42 %) is disproportionate to the percentage of injuries due to falls (19 %). Apart from fall from height, tripping, staggering, slipping caused a high proportion of accidents in reinforcement work (Memarian and Mitropoulos 2013a). Zhou et al. (2012) find that collapse accidents account for 51 % of all.

3.6.5.2 Hit by Objects, Mechanical Injuries and Electrocution Accidents

Immediate direct causes such as mechanical hazards (i.e. struck by hand tools, machinese, vehicles and cutting edges etc.), hit by falling objects (Ale et al. 2008; Irumba 2014; Zhou et al. 2012) and struck by objects (Memarian and Mitropoulos 2013a; Hinze et al. 1998). A cross-sectional survey of 201 large-size building projects, building records from Kampala City Council, safety statistics from the Department of Occupational Safety and Health, and accident investigation reports by Kampala City Council in 2008 shows that hit by objects and mechanical hazards, such as struck by hand tools, cutting edges, machines, vehicles etc. account for a large number of accidents on sites (Irumba 2014). Pratt et al. (1997)'s study shows that excavating machinery, tractors and cranes were the machines that the most frequently involved. The most common incident were struck by a mobile machine and boom. Weil (2001) agree that there is a high proportion of injuries (31 %) rise from workers being struck against or by an object. Memarian and Mitropoulos (2013b)'s study, in the same vein, find that overexertion, contact with and struck by object as the events with the highest frequency that lead to accident.

Electrocution is another major cause of accident that has been observed from previous literature (Zhou et al. 2012). For example, Suárez-Cebador et al. (2014)'s findings show that electrical accidents are almost five times more likely which lead to serious consequences than the average accident in the sector. Chi and Han (2013) show that 86 % of electrical shocks are the result of electrical wiring or apparatus. Pinto (2014) contact with electricity is another major causes of injury alongside injury caused by falling/swinging objects, hit by objects, contact with explosion, machinery or fire. Li and Bai (2008) find that head-on is the major causes for fatal accidents, rear-end was the major reason for dominant injury accident. A large percent of fatal accidents involved trucks whereas a large number of injuries involved light-duty vehicles. Many of the fatal accidents happen because of alcohol impairment, speeding, ignore the traffic control and speeding. On the other hand, followed too close caused high percentage of injury accidents. Weil (2001) records that although 18 % of deaths on site arise from vehicle related causes, less than 5 % of OSHA standards cited in that year are related to motor vehicles.

3.7 Conclusions

Construction accidents have often led to huge costs in compensation. Thus, the need to know the causes of accidents has become an important issue. The popularity of World Wide Web has not only bought convenience to companies, it has also opened the door to academic researchers from all over the World to obtain digital information. This chapter reviews that the digital information about the causes of accidents are mainly available in developed countries such as Europe and the US from the 5 academic journal database. Many researchers study the statistics obtained from official database or report, others obtain the data from questionnaire surveys or interviews. Despite the previous research proposes that triangulation, i.e. the usage of more than one research methods, is good to overcome the short-comings of the research method. Third, Science Direct and ASCE library contain more relevant journal articles than the other databases. Lastly, quite a large volume of research focus on the immediate causes of accidents, such as fall, slip and struck by objects and some of the workers' characteristics as their objective of study/findings of the results.

References

Adam, J. M., Pallarés, F. J., & Calderón, P. A. (2009). Falls from height during the floor slab formwork of buildings: Current situation in Spain. *Journal of Safety Research, 40*, 293–299.

Ahmad, N. H., & Seet, P. S. (2009). Dissecting behaviours associated with business failure: A qualitative study of SME owners in Malaysia and Australia. *Asian Social Science, 5*, 98–104.

Al-Htaybat, K. A. (2011). E-readiness framework for corporate online reporting. *Asian Social Science, 7*, 68–78.

Al-Rawajfih, K., Fong, S. F., & Idros, S. N. S. (2010). Stages of concern in integrating E-learning in discovery schools. *Asian Social Science, 6*, 54.

Ale, B. J. M., Bellamy, L. J., Baksteen, H., Damen, M., Goossens, L. H. J., Hale, A. R., et al. (2008). Accidents in the construction industry in the Netherlands: An analysis of accident reports using storybuilder. *Reliability Engineering & System Safety, 93*, 1523–1533.

Arboleda, C. A., & Abraham, D. M. (2004). Fatalities in trenching operations—analysis using models of accident causation. *Journal of Construction Engineering and Management, 130*, 273–280.

Arquillos, A. L., Romero, J. C. R., & Gibb, A. (2012). Analysis of construction accidents in Spain, 2003–2008. *Journal of Safety Research, 43*, 381–388.

Awang, M. M., Jindal-Snape, D., & Barber, T. (2013). A documentary analysis of the government's circulars on positive behavior enhancement strategies. *Asian Social Science, 9*, 203–208.

Cameron, I., Hare, B., & Davies, R. (2008). Fatal and major construction accidents: A comparison between Scotland and the rest of Great Britain. *Safety Science, 46*, 692–708.

Chan, M. (2011). Fatigue: The most critical accident risk in oil and gas construction. *Construction Management and Economics, 29*, 341–353.

Cheng, C. W., Leu, S. S., Cheng, Y. M., Wu, T. C., & Lin, C. C. (2012). Applying data mining techniques to explore factors contributing to occupational injuries in Taiwan's construction industry. *Accident Analysis and Prevention, 48*, 214–222.

Cheng, C. W., Lin, C.-C., & Leu, S.-S. (2010). Use of association rules to explore cause–effect relationships in occupational accidents in the Taiwan construction industry. *Safety Science, 48*, 436–444.

Chi, S., & Han, S. (2013). Analyses of systems theory for construction accident prevention with specific reference to OSHA accident reports. *International Journal of Project Management, 31*, 1027–1041.

Das, D. K. (2010). Globalisation: Past and present. *Economic Affairs, 30*, 66–70.

Dester, W., & Blockley, D. I. (1995). Safety—behaviour and culture in construction. *Engineering, Construction and Architectural Management, 2*, 17–26.

Dewlaney, K. S., Hallowell, M. R., & Fortunato III, B. R. (2012). Safety risk quantification for high performance sustainable building construction. *Journal of Construction Engineering and Management, 138*, 964–971.

Er, A. C. (2011). A comparative analysis of the Brazilian bioethanol sector and the Malaysian palm biofuel sector. *Asian Social Science, 7*, 74–78.

Frazier, M., & Ewing, S. (2009). Online free markets and entrepreneurship in the muslim world. *Economic Affairs, 29*, 28–35.

Glazner, J. E., Borgerding, J., Bondy, J., Lowery, J. T., Lezotte, D. C., & Kreiss, K. (1999). Contractor safety practices and injury rates in construction of the Denver International Airport. *American Journal of Industrial Medicine, 35*, 175–185.

Hale, A., Walker, D., Walters, N., & Bolt, H. (2012). Developing the understanding of underlying causes of construction fatal accidents. *Safety Science, 50*, 2020–2027.

Haslam, R. A., Hide, S. A., Gibb, A. G. F., Gyi, D. E., Pavitt, T., Atkinson, S., et al. (2005). Contributing factors in construction accidents. *Applied Ergonomics, 36*, 401–415.

Hinze, J., Pedersen, C., & Fredley, J. (1998). Identifying root causes of construction injuries. *Journal of Construction Engineering and Management, 124*, 67–71.

Hinze, J. W., & Teizer, J. (2011). Visibility-related fatalities related to construction equipment. *Safety Science, 49*, 709–718.

Hon, C. K. H., Chan, A. P. C., & Wong, F. K. W. (2010). An analysis for the causes of accidents of repair, maintenance, alteration and addition works in Hong Kong. *Safety Science, 48*, 894–901.

Hu, K., Rahmandad, H., Smith-Jackson, T., & Winchester, W. (2011). Factors influencing the risk of falls in the construction industry: A review of the evidence. *Construction Management and Economics, 29*, 397–416.

Huang, X., & Hinze, J. (2003). Analysis of construction worker fall accidents. *Journal of Construction Engineering and Management, 129*, 262–271.

Irumba, R. (2014). Spatial analysis of construction accidents in Kampala, Uganda. *Safety Science, 64*, 109–120.

Jabar, M. A., Usman, U. A., & Sidi, F. (2014). Usability evaluation of universities' websites. *International Journal of Information Processing and Management, 5*, 10–17.

Kerdvibulvech, C., & Yin, L. (2014). A new online database system based on virtually digital mapping. *International Journal of Information Processing and Management, 5*, 14–24.

Kirk, C. P., Chiagouris, L., Gopalakrishna, P. (2012). Some people just want to read: the roles of age, interactivity, and perceived usefulness of print in the consumption of digital information products. *Journal Of Retailing and Consumer Services, 19*, 168–178.

Lambert, R. (2007). The future of the news in the digital era. *Economic Affairs, 27*, 56–63.
Langford, D., Rowlinson, S., & Sawacha, E. (2000). Safety behaviour and safety management: Its influence on the attitudes of workers in the UK construction industry. *Engineering, Construction and Architectural Management, 7*, 133–140.
Laufer, A. (1987). Construction accident cost and management safety motivation. *Journal of Occupational Accidents, 8*, 295–315.
Leu, S. S., & Chang, C. M. (2013). Bayesian-network-based safety risk assessment for steel construction projects. *Accident Analysis & Prevention, 54*, 122–133.
Li, R. Y. M. (2013). The usage of automation system in smart home to provide a sustainable indoor environment: A content analysis in web 1.0. *International Journal of Smart Home, 7*, 47–59.
Li, R. Y. M., & Poon, S. W. (2007). A critical review of construction safety motivation in Hong Kong. In *Proceedings of the Management Science and Engineering Academic Conference* (pp. 13–19). Tianjin: Tianjin University.
Li, R. Y. M., & Poon, S. W. (2009). Future motivation in construction safety knowledge sharing by means of information technology in Hong Kong. *Journal of Applied Economic Sciences, 4*, 457–472.
Li, R. Y. M., & Poon, S. W. (2013). *Construction safety*. Germany: Springer.
Li, Y., & Bai, Y. (2008). Comparison of characteristics between fatal and injury accidents in the highway construction zones. *Safety Science, 46*, 646–660.
Liao, C. W. (2012). Pattern analysis of seasonal variation in occupational accidents in the construction industry. *Procedia Engineering, 29*, 3240–3244.
Lin, Y. H., Chen, C. Y., & Wang, T. W. (2011). Fatal occupational falls in the Taiwan construction industry. *Journal of the Chinese Institute of Industrial Engineers, 28*, 586–596.
Lópeza, M. A. C., Fontanedab, I., Alcántarab, O. J. G., & Ritzel, D. O. (2011). The special severity of occupational accidents in the afternoon: The lunch effect. *Accident Analysis and Prevention, 43*, 1104–1116.
Lu, X., & Cao, Y. (2013). Digital information encryption using multiple Fourier transforms and decimal expansion of irrational numbers. *Optik, 124*, 1202–1206.
Memarian, B., & Mitropoulos, P. (2013a). Accidents and minor accidents of the musculoskeletal system in heavy (concrete reinforcement work) and light (painting) construction work. *Safety Science, 59*, 179–186.
Memarian, B., & Mitropoulos, P. (2013b). Accidents in masonry construction: The contribution of production activities to accidents, and the effect on different worker groups. *Safety Science, 59*, 179–186.
Nelson, N. A., Kaufman, J., Kalat, J., & Silverstein, B. (1997). Falls in construction: Injury rates for OSHA-inspected employers before and after citation for violating the Washington state fall protection standard. *American Journal of Industrial Medicine, 31*, 296–302.
Perez-Alonso, J., Carreno-Ortega, A., Vazquez-Cabrera, F. J., & Callejon-Ferre, A. J. (2012). Accidents in the greenhouse-construction industry of SE Spain. *Applied Ergonomics, 43*, 69–80.
Pinto, A. (2014). QRAM a qualitative occupational safety risk assessment model for the construction industry that incorporate uncertainties by the use of fuzzy sets. *Safety Science, 63*, 57–76.
Pratt, S. G., Kisner, S. M., & Moore, P. H. (1997). Machinery-related fatalities in the construction industry. *American Journal of Industrial Medicine, 32*, 42–50.
Rowlinson, S., Mohamed, S., & Lam, S. W. (2003). Hong Kong construction foremen's safety responsibilities: A case study of management oversight. *Engineering, Construction and Architectural Management, 10*, 27–35.
Saram, D. D. D., & Tang, S. L. (2005). Pain and suffering costs of persons in construction accidents: Hong Kong experience. *Construction Management and Economics, 23*, 645–658.
Seo, M. S., & Park, D. W. (2014). A study on the encryption of searching, deleting, and restricting access of personal information in smart PCs. *International Journal of Information Processing and Management, 5*, 124–131.

Suárez-Cebador, M., Rubio-Romero, J. C., & López-Arquillos, A. (2014). Severity of electrical accidents in the construction industry in Spain. *Journal of Safety Research, 48*, 63–70.
Sui, M., Li, J., Dong, C., & Bai, Y. (2013). The encrypted watermarking for medical image based on Arnold scrambling and DWT. *Journal of Convergence Information Technology, 8*, 893–902.
Suraji, A., Duff, A. R., & Peckitt, S. J. (2001). Development of causal model of construction accident causatgion. *Journal of Construction Engineering and Management, 127*, 337–344.
Weil, D. (2001). Assessing OSHA performance: New evidence from the construction industry. *Journal of Policy Analysis and Management, 20*, 651–674.
Wu, W., Yang, H., Chew, D. A. S., Yang, S. H., Gibb, A. G. F., & Li, Q. (2010). Towards an autonomous real-time tracking system of near-miss accidents on construction sites. *Automation in Construction, 19*, 134–141.
Zhang, X. (2011). Talk about the change and development of college library since the 30 years of reform and opening-up. *Asian Social Science, 7*, 219–221.
Zhou, Z., Li, Q., & Wu, W. (2012). Developing a versatile subway construction incident database for safety management. *Journal of Construction Engineering and Management, 138*, 1169–1180.
Zou, P. X. W., & Zhang, G. (2009). Comparative study on the perception of construction safety risks in China and Australia. *Journal of Construction Engineering and Management, 135*, 620–627.

Chapter 4
Relevant Data in the Rising Tide of Big Data: A Text-Mining Analysis in Construction Safety Index

Rita Yi Man Li

Abstract The previous generation of academia faced the problem of insufficient data, academic journal articles and books. Thanks to the rapid development of World Wide Web since 90s, academic papers published in one place can be viewed in another side of the globe, insufficient literature and data problems have been relieved. Modern academic researchers, however, face another major problem of big data. There are too many irrelevant articles. Much of the time has been spent to search for the relevant articles from the irrelevant ones. One of the major aims of this chapter is to review the problem of big data based on the example of construction safety index. Construction safety is an important issue Worldwide. The development of construction safety index is of particular importance as it provides an objective measurement on the level of safety on sites. This chapter reviews (1) how intelligence is the academic search engine in screening the relevant articles out the irrelevant materials, (2) compare and contrast the methods which are used to construct the construction safety indices with the other indices constructed worldwide.

Keywords Construction safety index · Content analysis · Academic article visibility · Academic journal database

4.1 Introduction

As the number of citations per article is often treated as one of the important indicators in research output quality (Ebrahim et al. 2013) and the articles have to be known to the other academics before the articles are cited, researchers in

R.Y.M. Li (✉)
Department of Economics and Finance / Sustainable Real Estate Research Center,
Hong Kong Shue Yan University, Hong Kong, China
e-mail: ritarec1@yahoo.com.hk

R.Y.M. Li
School of Natural and Built Environments, University of South Australia, Adelaide, Australia

R.Y.M. Li, *Construction Safety and Waste Management*,
Risk Engineering, DOI 10.1007/978-3-319-12430-8_4

academia realizes the importance of article visibility. Gone are the days when academic writers do their research in libraries, databases which hosts millions of academic research papers online play a more important role in modern era. Likewise, article visibility no longer depends on librarian, but by the keyword search in database. By inserting the keywords to the search query box, however, may return with thousands or more than ten thousands of irrelevant works. Whilst the academia in the olden days suffer from the problems of too few data, modern academia suffer from the problem of big data—with too much irrelevant data but few are relevant works. This chapter reviews the problem of big data in construction safety index research. It includes 6 academic databases in City University of Hong Kong: Emerald Management Xtra, Science Direct, Taylor and Francis Online, EBSCO, Proquest and Springerlink. In the following section, it firstly reviews the importance of academic article visibility, follows by an analysis of different types of indices' construction methods and finally the construction index construction methods which are found in the academic journal database.

4.2 The Importance of Academic Article Visibility

To publish one article, the researcher used approximately 18 citations on average (Saechan 2014). In academia, number of citations per academic staff is often used as a criteria to rank the university Worldwide. In Times Higher Education, citations impact count 30 % among all the factors which include teaching and learning environment, research volume, international staff and income from industry (Times Higher Education 2014). Similar considerations can also be found in QS University ranking method. Citations count for 20 % (QS 2013). This rule of game has extended to academic staff's promotion in higher education institutions.

Nevertheless, how to let the others to know the existence of the piece of research is an important issue before getting cited in big data or knowledge explosion era. "If we ever think of our knowledge base expansion in print publication only, it is reasonable to imagine a world of text which is sufficient to cover our globe to a depth of 1 in. up to 1,850. Hundreds of years later, amount of printed materials have doubled to a depth of 2 in. By 1950, they have expanded to a depth of 36 in." (Li 2012). While publishing a high quality paper in scientific journals will be a halfway of receiving citation in the future, the rest depends on "advertising" and disseminating the publications with proper research tools. The more researchers and students in other fields notice the presence of the research article, the more citations they will have in the future. Some academic researchers may even argue that citations of an article might depend more on visibility than the merit of the article. High visibility article tends to have more downloads and later citations records. Hence, another of the major way to increase citations is to expose the research output to the widest possible academic audience (Ebrahim et al. 2013). All these lead the flourish of various free academic online network such as Research Gate, Academia. Some universities such as the University of Hong Kong has launched

Table 4.1 University ranking methodology used by Times Higher Education (2014) and QS (2013)

Criterion	Indicator used by Times	Weighting used by Times' ranking method (%)	Indicator used by QS	Weighting used by QS (%)
Citations—research influence	Citations impact (average citations per paper according to Thomson Reuter's Web of Science)	30	Citations per faculty as an indication of the research impact with the data obtained from Scopus	20
Teaching—the learning environment	Teaching reputation survey	30	Faculty to student ratio is used to indicate the commitment to teaching	20
	Income earned per academic Ph.D. or bachelor's awards			
	Undergraduates admitted per academic			
	Ph.D. awards per academic			
Research—volume, income and reputation	Papers written by academic and research staff	30	Nil	0
	Research income			
	Reputation survey about research			
International mix—staff and students	Ratio of international to domestic staff	7.5	International student ratio which measures the diversification of international students' community	
	Ratio of international to domestic students			
	Proportion of international co-authored research papers			
Industry	Research income obtained from industry per academic staff	2.5	Employer reputation is quantified according to a global survey of graduates' employers	0
Academic reputation	Nil	0	Academic reputation according to a global survey of academics	0

the online scholar hub to allow us to download the academic staff's articles freely. It also leads to a research question that is suggested by the author, how easy it is to find some relevant articles out of the irrelevant ones by using keywords search in traditional academic databases available in the library online (Table 4.1).

4.3 The Rise and Problem of Big Data

The growth of the abovementioned academic databases eventually turns the traditional close access articles to de facto "open access". All the academic researchers can upload and download some academic articles online. Outside the academic circle, the birth of free and easy to use website making tools such as Wordpress, some Web 2.0 tools such as YouTube and online social network such as Facebook, LinkedIn, Google+ etc. essentially speed up the exponential growth in volume of data. Beyond doubt, online databases are gaining weighs as avenues for publishing data. All these, however, also implies that revolutionary measures are necessary for data accessibility, management and analysis (Howe et al. 2008). How to link up the data as efficiently as human brain may be an issue as pointed out as the post Web 2.0 era or the so-called Web 3.0 epoch. For example, Chu and Yang (2012) incubates the establishment of semantic web services for next generation electronic library via Web 3.0, a kind of semantic social web mechanisms. Unlike Web 2.0 which allows the readers to read, edit and rewrite the content online (Li 2011; Li and Ah Pak 2010), it is foreseen that Web 3.0 will be more intelligent than Web 2.0 which integrate Distributed Artificial Intelligence, semantics web and intelligent agent into the ubiquitous networks (Chu and Yang 2012).

4.4 Indices Construction Method

To quantify the society's phenomenon, such as the stock prices movement, and to compare the different level of human development, we have developed different indices for business and investment purposes. For example, Heng Seng Index is used to measure the movement of a basket of stock prices in Hong Kong, World development index is used to compare different level of development Worldwide. Likewise, there are tons of different ways to construct various indices in academia. Table 4.2 shows the indices that are related to construction project (e.g. Xu and Moon 2013; Hassanein and Khalil 2008), housing price (e.g. Wu et al. 2012), consumer (Koster et al. 2014) and environment (e.g. Messer et al. 2014). Among all of them, quite a number of the research articles adopt econometric techniques. For example, Xu and Moon (2013) construct the construction costs index with the Cointegrated Vector Autoregression Model, Wu et al. (2012) and Dusan and Victor (2011) adopts the Hedonic Model to estimate the housing price and cost of living index. Another group of the indices are constructed with statistical modelling, for

Table 4.2 The construction methods of the indicies (author's Table)

Author	Name of the index	Index construction method
Du et al. (2014)	Financial center index	FC-niche formation process
Dusan and Victor (2011)	Cost of living index	Hedonic model with substitution bias considered
Hajkowicz (2006)	Multi-attributed index	Multi-attributed utility theory
MacDorman et al. (2007)	Acoustic signal to emotion index	Pleasure, arousal and dominance (PAD) model
Wu et al. (2012)	Housing price index in the nascent market	Simple average method, matching approach with repeating sales modeling framework, hedonic modeling approach
Gündüz et al. (2013)	Relative importance index for construction projects in Turkey	Relative importance index method (ranking importance among factors)
Xu and Moon (2013)	Construction cost index	Cointegrated vector autoregression model
Shahandashti and Ashuri (2013)	Construction cost index	Vector error correction model
Kodric and Lea (2012)	Index of capital services on construction price	Perpetual inventory method
Hassanein and Khalil (2008)	Construction cost index	Ordinary least squares method and forecasting
Wang and Mei (1998)	Forecasting of construction cost index	Weighted average
Messer et al. (2014)	Environmental quality index	Principal components analysis (PCA)
Dyer et al. (2012)	Construction inflation index	Bayesian information criteria
Yadav et al. (2011)	Cost escalation in construction	Moving average method
Koster et al. (2014)	Consumer quality (CQ) index	Factor analysis
Sylvie et al. (2013)	Construction security rating index	Security vulnerability assessment (SVA) method
Yu and Ive (2008)	Building price index	Hedonic price model
Eom and Paek (2009)	Environmental disputes risk index	Analytic hierarchy process model
Yeung et al. (2007)	Partnering performance index (PPI)	Delphi survey
Minchin et al. (2008)	Construction quality index	Analytic hierarchy process model
Li and Liu (2010)	Malmquist index	Malmquist (Bilateral) index method with a novel decomposition technique

(continued)

Table 4.2 (continued)

Author	Name of the index	Index construction method
Hamdi et al. (2014)	International roughness index	Regression
Ratick and Osleeb (2013)	Vulnerability indices	Weighted average, ordered weighted average, data envelopment analysis
Barrios and Komoto (2006)	Sustainable development index	Sparse principal components analysis (SPCA)
Antonio and Jesus (2011)	Multidimensional well-being index for the Spanish regions	Principal components analysis (PCA)
Hardjosoekarto (2012)	Social development index	SSM-based AR (action Research)
Yanuar et al. (2010)	Health index	Structural equation modeling
Bernard (2004)	Social vulnerability index	SWOT analysis

example, Du et al. (2014) uses the FC-niche Formation Process to formulate the financial center index. Messer et al. (2014) uses of the Principal Components Analysis (PCA) to construct the environmental quality index and Barrios and Komoto (2006) uses the Sparse Principal Components Analysis (SPCA) to construct the sustainable development index.

4.5 Research Method

As the volume of publication and the underlying knowledge base is expanding at an increasing rate. Texting mining plays an important role in this aspect. Text classification attempts to determine whether a document has particular characteristics of interest with regards to a given topic (Cohen and Hersh 2005). Text mining uses the data mining techniques, information retrieve and extraction processes to provide a more structured analysis of textual knowledge. The results obtained are more meaningful than simple words search (Ananiadou et al. 2006). Text mining has been used in different types of research. For example, Driel et al. (2006) uses text mining to classify more than 5,000 human phenotypes in the Online Mendelian Inheritance's Man database. In patent analysis, text mining is used as information-extracting and data-processing tool to change the raw data to structured data (Yoon and Park 2004).

In the same vein, the research method for this chapter is text mining. With the keyword search "construction safety index" and construction safety index without quotation, the methods which are used to construct the construction safety indices Worldwide are searched via the major academic journal database such as Science Direct, Emerald Management Xtra, Taylor and Francis Online, Springerlink, EBSCO and Proquest.

4.6 Research Results and Discussion

The research results show that although many of the databases list out a huge number of the so-called relevant articles. By the time the author goes through each of them, the relevant articles about construction safety indices are scarce. For example, while there are 45,141 articles are claimed to be relevant about the construction safety in Science Direct, there are only 7 of them are relevant when the authors read them bit by bit. Similarly the screen shows that there are 33,377 relevant articles in Springerlink, only 1 is relevant. All these sheds light on the problems of big data. The research results, however, open a new research agenda for future construction safety research:

1. Majority of the construction index studies are built for the construction industry in China. The author speculate that the reasons for this is:

    i. There is a fast economic growth in the urban areas, leading to fast pace of urban-rural migration, a lot of high-rise buildings are built in mainland China, leading to a higher number of construction accidents as well as the need to construct the construction safety index;

    ii. There is an increasing need for new infrastructure such as train, bridges and roads. As the construction process is complex and lots of accidents happen on sites, all these lead to an urgent needs to locate some people, management styles etc. which may lead to higher accident rates.

2. Previous construction safety index mainly focuses on safety management or some specific worksites such as tower crane, metro, highway, hydropower plant

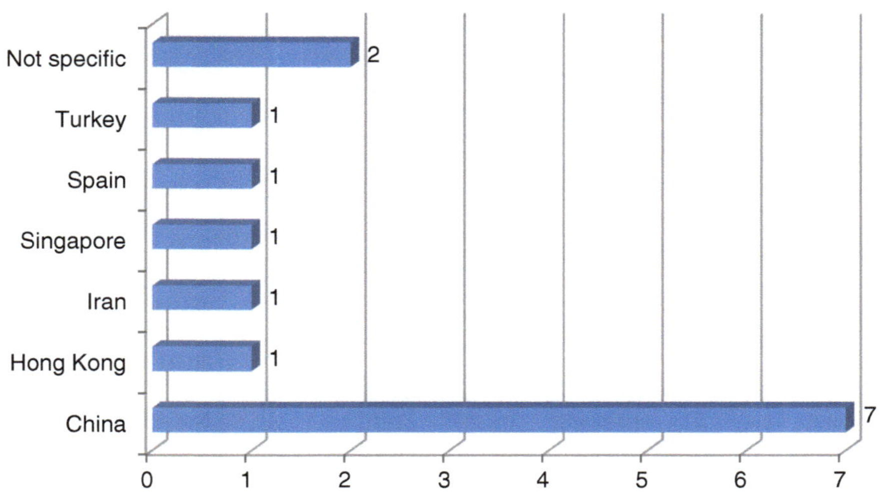

Fig. 4.1 Number of articles that studies the construction safety index/indicies in different cities or countries (author's figure)

but none of them shed light on towers of building work. This reflects the fact that many buildings in cities such as Adelaide are low-rise. There may be more accidents in more complicated and longer construction projects. For example, housing here may be one storey only, the likelihood of serious accidents may not the likely happen as compared to civil engineering work like bridges and tunnels. There is a more urgent needs for construction indices among these complicated works with more accidents happen on sites.

3. In spite of the previous research shows that there are multiple reasons which lead to construction accidents, few have studied the human errors but none of the index has included workers' background such as safety education and training and the four aspects in constructing construction safety index (Fig. 4.1, Tables 4.3 and 4.4).

Table 4.3 Research results of the academic journal databases

Database	Keyword use	Number of articles show on the screen	Number of relevant article (construction safety index)	Percentage of relevant articles as compared to the articles show in the screen (%)
Science Direct	"Construction safety index"	1	0	0.00
Science Direct	Construction safety index	45,141	7	0.02
Emerald Management Xtra	"Construction safety index"	2	0	0.00
Emerald Management Xtra	Construction safety index	1,564	0	0.00
Taylor and Francis Online	"Construction safety index"	0	0	Not applicable
Taylor and Francis Online	Construction safety index	0	0	Not applicable
Springerlink	"Construction safety index"	0	0	Not applicable
Springerlink	Construction safety index	33,377	1	0.00
EBSCO	"Construction safety index"	1	1	100.00
EBSCO	Construction safety index	10	3	30.00
Proquest database (full text, peer reviewed, scholarly journal)	"Construction safety index"	0	0	Not applicable
Total number of articles		80,096	12	0.01

Table 4.4 Research results of construction safety indices

Author(s)	Aims of the construction safety indices	Countries/Cities	Types of construction work	Research method
Zhou et al. (2014)	To investigate human error in construction accidents	China	Large-scale hydropower-construction projects	The human factor analysis and classification system (HFACS)
Zhang et al. (2014)	To reveal the critical basic events which help reduce the risk in metro construction	China	Metro construction	Fault tree analysis
Reyes et al. (2014)	To develop the Health and Safety Costs Index according to economic criteria and quantify economic terms	Spain	Industrial premises construction	An exponential function
Liu et al. (2012)	To construct construction safety index based on (1) safety product management system, (2) qualification, institutions and personnel management, (3) safety target, safety technology, equipment and facilities management	China	Not specific	Fuzzy-Entropy Theory (with experts' safety inspection)
Chen and Tian (2012)	Safety index is used to evaluate safety behavior and change in trend in presence of behavior based safety	China	Construction sites which adopt behavior based safety management method	The equation: SI = 100[N2/(N1 + N2)]% where N2: observed times of safety behavior; N1: observed times of unsafe behavior; N1 + N2: total times of observed behavior
Wu and Shen (2012)	To construct the construction safety evaluation index system	China	Highway construction sites	Fuzzy grey relational analysis theory and AHP
Hon et al. (2014)	It adopts the safety climate index developed by the Occupational Safety and Health Council (OSHC)	Hong Kong	RMAA work	Simply adopts the index developed by OSHC
Zheng et al. (2012)	AHP Safety index is built to reflect the safety condition resulting from hot and humid climate	N/A	N/A	AHP

(continued)

Table 4.4 (continued)

Author(s)	Aims of the construction safety indices	Countries/ Cities	Types of construction work	Research method
Zhou and Kou (2010)	Failure likelihood index method and analytic hierarchy process is used to estimate human error probabilities	China	Structural failures	AHP and failure likelihood index method
Teo and Ling (2006)	This safety index is developed to test the effectiveness of safety management system	Singapore	Not specific	AHP
Shapira and Simcha (2009)	To provide an assessment on tower cranes' relative important safety factors			AHP and interviews
Fang et al. (2004)	The safety management index is used as a means to evaluate real-time safety management performance by using key management factors	China	Not specific	Survey and interview

4.7 Conclusion

Previous research on the methods which are used to construct the construction safety index Worldwide via searching through the major academic journal database such as Science Direct, Emerald Management Xtra, Taylor and Francis Online, Springerlink, EBSCO and Proquest. Similar to any other economic, social and environmental indices, our research shows that the development of construction index adopts statistical tools such as Fault Tree Analysis and AHP. Unlike the construction of economic indices, some of the construction indices are constructed according to the results of interviews. This chapter also illustrates that current relevant research on construction indices are quite scarce to when we use the keyword construction safety index even though the database shows that there are more than 10,000 relevant results. Majority are irrelevant. It showcases the problem of big data in the World Wide Web.

Acknowledgments The author would like to thank Mr T.H. Leung's research assistance.

References

Ananiadou, S., Kell, D. B., & Tsujii, J.-I. (2006). Text mining and its potential applications in systems biology. *Trends in Biotechnology, 24*, 571–579.

Antonio, J., & Jesus, P. M. (2011). Construction and evolution of a multidimensional well-being index for the Spanish regions. *Social Indicators Research, 107*, 259–279.

Barrios, E., & Komoto, K. (2006). Some approaches to the construction of a sustainable development index for the Philippines. *International Journal of Sustainable Development and World Ecology, 13*, 277–288.

Bernard, G. S. (2004). Toward the construction of a social vulnerability index—theoretical and methodoligical considerations. *Social and Economic Studies, 53*, 1–29.

Chen, D., & Tian, H. (2012). Behavior based safety for accidents prevention and positive study in China construction project. *Procedia Engineering, 43*, 528–534.

Chu, H.-C., & Yang, S. W. (2012). Innovative semantic web services for next generation academic electronic library via web 3.0 via distributed artificial intelligence. *Intelligent Information and Database Systems Lecture Notes in Computer Science, 7196*, 118–124.

Cohen, A. M., & Hersh, W. R. (2005). A survey of current work in biomedical text mining. *Breifings in Bioinformatics, 6*, 57–71.

Driel, M. A. V., Bruggeman, J., Vriend, G., Brunner, H. G., & Leunissen, J. A. M. (2006). A text-mining analysis of the human phenome. *European Journal of Human Genetics, 14*, 535–542.

Du, H., Xia, Q., Ma, X., & Chai, L. (2014). A new statistical dynamic analysis of ecological niches for China's financial centres. *Physica A: Statistical Mechanics and its Applications, 395*, 476–486.

Dusan, P. A., & Victor, I. R. (2011). Subtitution bias and the construction of a spatial cost of living index. *Papers in Regional Science, 92*, 103–117.

Dyer, B., Goodrum, P. M., & Viele, K. (2012). Effects of omitted bias on construction real output and its implications on productivity trends in the United States. *Journal of Construction Engineering and Management, 138*, 558–566.

Ebrahim, N. A., Salehi, H., Embi, M. A., Tanha, F. H., Gholizadeh, H., Motahar, S. M., et al. (2013). Effective strategies for increasing citation frequency. *International Education Studies, 6*, 93–99.

Eom, S. J., & Paek, J. H. (2009). Risk index model for minimizing environmental disputes in construction. *Journal of Construction Engineering and Management, 135*, 34–41.
Fang, D. P., Huang, X. Y., & Hinze, J. (2004). Benchmarking studies on construction safety management in China. *Journal of Construction Engineering and Management, 130*, 424–432.
Gündüz, M., Nielsen, Y., & Özdemir, M. (2013). Quantification of delay factors using the relative importance index method for construction projects in Turkey. *Journal of Management in Engineering, 29*, 133–139.
Hajkowicz, S. (2006). Multi-attributed environmental index construction. *Ecological Economics, 57*, 122–139.
Hamdi, A. S., Tighe, S. L., & Li, N. Y. (2014). Canadian calibration on mechanistic—empirical pavement design guide to estimate International Roughness Index (IRI) using MTO Data. *International Journal of Pavement Research and Technology, 7*, 101–108.
Hardjosoekarto, S. (2012). Construction of social development index as a theoretical research practice in action research by using soft systems methodology. *Systemic Practice and Action Research, 25*, 493–509.
Hassanein, A. G., & Khalil, N. L. (2008). Building Egypt 1—a general indicator cost index for the Egyptian construction industry. *Cost Engineering, 50*, 18–26.
Hon, C. K. H., Chan, A. P. C., & Yam, M. C. H. (2014). Relationships between safety climate and safety performance of building repair, maintenance, minor alteration, and addition (RMAA) works. *Safety Science, 65*, 10–19.
Howe, D., Costanzo, M., Fey, P., Gojobori, T., & Hannick, L. (2008). Big data: The future of biocuration. *Nature, 455*, 47–50.
Kodric, B., & Lea, B. (2012). The sensitivity of capital services index on construction price indices. *Economic Research, 25*, 666–689.
Koster, E. B., Ong, R. S., Heybroek, R., Delnoij, M. J., & Baars, E. W. (2014). The consumer quality index anthroposophic healthcare: A construction and validation study. *BMC Health Services Research, 14*, 1–22.
Li, R. Y. M. (2011). Game theory analysis on underground space planning knowledge sharing via Web 2.0. In: *5th International Conference on New Trends in Information Science and Service Science*, Macao, 2011.
Li, R. Y. M. (2012). Knowledge management, sharing and creation in developing countries' banking industries. *Advanced in Network and Communications, 1*, 13–26.
Li, R. Y. M., & AH PAK, D. H. (2010). Resistance and motivation to share sustainable development knowledge by Web 2.0. *Journal of Information and Knowledge Management, 9*, 251–262.
Li, Y., & Liu, C. L. (2010). Malmquist indices of total factor productivity changes in the Australian construction industry. *Construction Management and Economics, 28*, 933–945.
Liu, D., Yu, L., & Li, B. (2012). Fuzzy-entropy theory comprehensive evaluation method and its application in building construction safety. *Procedia Engineering, 43*, 137–142.
Macdorman, K. F., Ough, S., & Ho, C. C. (2007). Automatic emotion prediction of song excerpts: Index construction, algorithm design, and empirical comparison. *Journal of New Musical Research, 36*, 283–301.
Messer, L. C., Jagai, J. S., Rapppazzo, K. M., & Lobdell, D. T. (2014). Construction of an environmental quality for public health research. *Environemental Health, 13*, 1–39.
Minchin, R. E., Hammons, M. I., & Ahn, J. Y. (2008). A construction quality index for highway construction. *Construction Management and Economics, 26*, 1313–1324.
QS. (2013). *QS world university rankings: Methodology*. Retrieved September 5, 2014, from http://www.topuniversities.com/university-rankings-articles/world-university-rankings/qs-world-university-rankings-methodology.
Ratick, S. J., & Osleeb, J. P. (2013). Measuring the vulnerability of populations susceptible to lead contamination in the Dominican Republic: Evaluating composite index construction methods. *GeoJournal, 78*, 259–272.
Reyes, J. P., San-Jos, J. T., Cuadrado, J., & Sancibrian, R. (2014). Health and safety criteria for determining the sustainable value of construction projects. *Safety Science, 62*, 221–232.

Saechan, C. (2014). Patterns and problems of information use in the production of articles by academics in the humanities and social sciences in Thailand. *Asian Social Science, 10*, 115–127.

Shahandashti, S. M., & Ashuri, B. (2013). Forecasting engineering news-record construction cost index using multivariate time series models. *Journal of Construction Engineering and Management, 139*, 1237–1243.

Shapira, A., & Simcha, M. (2009). AHP-based weighting of factors affecting safety on construction sites with tower cranes. *Journal of Construction Engineering and Management, 135*, 307–318.

Sylvie, J. R., Thomas, S. R., Lee, S. H., Chapman, R. E., & Smith, R. T. (2013). Development and interpretation of the security rating index. *Journal of Construction Engineering and Management, 139*, 185–194.

Teo, E. A. L., & Ling, F. Y. Y. (2006). Developing a model to measure the effectiveness of safety management systems of construction sites. *Building and Environment, 41*, 1584–1592.

Times Higher Education. (2014). *Times higher education world university rankings*. Retrieved September 5, 2014, from http://www.universityrankings.ch/methodology/times.

Wang, C. H., & Mei, Y. H. (1998). Model for forecasting construction cost indices in Taiwan. *Construction Management and Economics, 16*, 147–157.

Wu, J., Deng, Y., & Liu, H. (2012). House price index construction in the nascent housing market: The case of China. *The Journal of Real Estate Finance and Economics, 48*, 522–545.

Wu, Z., & Shen, R. (2012). Safety evaluation model of highway construction based on fuzzy grey theory. *Procedia Engineering, 45*, 64–69.

Xu, J. W., & Moon, S. W. (2013). Stochastic forecast of construction cost index using a cointegrated vector autoregression model. *Journal of Management in Engineering, 29*, 10–18.

Yadav, S., Shedabale, K., Suman, S., & Tanga, D. (2011). Cost escalation in construction: An alternative approach. *IUP Journal of Infrastructure, 9*, 51–62.

Yanuar, F., Ibrahim, K., & Jemain, A. A. (2010). On the application of structural equation modeling for the construction of a health index. *Environmental Health and Preventive Medicine, 15*, 285–291.

Yeung, F. Y., Chan, P. C., Chan, W. M., & Li, L. K. (2007). Development of a partnering performance index (PPI) for construction projects in Hong Kong: A Delphi study. *Construction Management and Economics, 25*, 1219–1237.

Yoon, B., & Park, Y. (2004). A text-mining-based patent network: Analytical tool for high-technology trend. *Journal of High Technology Management Research, 15*, 37–50.

Yu, K. W., & Ive, G. (2008). The compilation methods of building price indices in Britain: A critical review. *Construction Management and Economics, 26*, 693–705.

Zhang, L., Skibniewski, M. J., Wu, X., Chen, Y., & Deng, Q. (2014). A probabilistic approach for safety risk analysis in metro construction. *Safety Science, 63*, 8–17.

Zheng, G., Zhu, N., Tian, Z., Chen, Y., & Sun, B. (2012). Application of a trapezoidal fuzzy AHP method for work safety evaluation and early warning rating of hot and humid environments. *Safety Science, 50*, 228–239.

Zhou, C., & Kou, X. J. (2010). Method of estimating human error probabilities in construction for structural reliability analysis based on analytic hierarchy process and failure likelihood index method. *Journal of Shanghai Jiaotong University (Science), 15*, 291–296.

Zhou, J. L., Bai, Z. H., & Sun, Z. Y. (2014). A hybrid approach for safety assessment in high-risk hydropower-construction-project work systems. *Safety Science, 64*, 163–172.

Chapter 5
An Institutional Economic Analysis on Construction Safety Knowledge Sharing and E-Learning via Mobile Apps

Rita Yi Man Li

Abstract The growth of World Wide Web and mobile technology such as mobile phones and tablets nurture the mobile learning development. Construction workers can now learn anytime and anywhere at affordable costs. Transaction costs of obtaining various information and knowledge can be reduced substantially with various mobile apps. Some workers who work alone can also obtain the knowledge about their jobs or first aid knowledge via wireless devices. Likewise, the popularity of various apps provide an alternative channels to construction firms and safety officers who wish to share the construction safety knowledge with construction workers who may have high time costs to attend safety lectures and tutorials. This chapter reviews the present mobile apps usage in safety knowledge sharing and m-learning feasibility. It is found that many of the safety mobile apps are used for (1) building safety inspection checklists, recording the safety meeting, safety accidents (2) knowledge sharing via videos and (3) calculate safety risks according to different types of indices such as heat and heavy lifting.

Keywords Knowledge sharing · E-Learning · Mobile apps

5.1 Introduction

The rapid expansion of the internet creates a golden opportunity to exchange, gather and disseminate information via World Wide Web (www). (Dutta et al. 2009). Web services utilize the internet as one large platform of Web protocols with open

---

R.Y.M. Li (✉)
Department of Economics and Finance / Sustainable Real Estate Research Center, Hong Kong Shue Yan University, Hong Kong, China
e-mail: ritarec1@yahoo.com.hk

R.Y.M. Li
School of Natural and Built Environments, University of South Australia, Adelaide, Australia

R.Y.M. Li, *Construction Safety and Waste Management*,
Risk Engineering, DOI 10.1007/978-3-319-12430-8_5

standards, such as SOAP, HTTP and XML to develop various information systems according to different programming languages which are used on heterogeneous platforms (Chang and Hsiao 2011). The high expandability and cross-platform information exchange of XML, for example, reduce the difficulties in exchanging resources between heterogeneous databases, resulting from new database architectures and formats (Chang and Hsiao 2011). The popularity of www also leads to an increasing need to study the behavior of Web-user with efficient Web-prediction technique Worldwide (Dutta et al. 2009). For example, Jansen et al. (2000) analyze transaction logs of 51,473 queries posed by 18,113 Excite, a major online search service users. O'Cass and Fenech (2003)'s findings indicate that Technology Acceptance Model provides good theoretical framework to understand Web users' adoption for retail purposes.

On the other hand, as web browsers rely on Information and Communication Technology (ICT) to view, edit and share the information in www, ICT has now become an important and indispensible tool in communication and learning. It also offers golden income generating opportunities, active participation in decision making processes, provides a viable tools which overcomes the limitations of time and space, knowledge and information sharing. All in all, it enhances efficiency and transparency at work and everyday life activities. In general, popular ICT tools include e-information processing tools such as desktop computers, notebook, mobile phones, various fixed-line telecommunication technologies, wireless communications and specialized application devices such as Global Positioning Systems and barcode scanners and Braille readers. All these digital technologies unlock the information that we need, increase productivity, enrich the quality of lives and innovation (Samah et al. 2010). The new generations of mobile phones allow us to view websites on line, send emails, join Whatsapps group anywhere and anytime have led to another breakthrough in a new form of e-learning, i.e. m-learning (Hrbáčková and Vávrová 2012). Without doubt, the rapid development in m-learning recent years changes the teaching and learning façade. This growth has been fastened by an advance development in mobile technology with an introduction of third generation wireless system allow high-speed web service over the internet, as well as the production of mobile devices that includes tablet PCs, multimedia-enabled smartphone, Samsung Galaxy Tab and Apple iPad (Norman et al. 2014). Furthermore, the popularity of the mobile apps imply that an app can be used by many people around the World, removing the geographical boundary. Modern learning is beyond traditional building envelope and the fixed internet. Indeed, the advancement of mobile technology opens a myriad of learning opportunities (Hussin et al. 2012).

Although the previous research finds that knowledge sharing among construction workforce is restricted in traditional face-to-face interaction and the use of information technology in construction companies is piecemeal and limited (Li and Poon 2009), the increasing population of generation Y and generation Z who are expert users in computers may lead to another revolution of m-learning, mobile learning and mobile safety knowledge sharing on site (Li 2012). This chapter aims to study whether the construction safety apps help construction safety knowledge sharing and e-learning.

5.2 Knowledge

Knowledge is a combination of experiences which are gained and accumulated from human practices during the World transformation. It is a summary and distillation of information. Previous literature suggests that it is an philosophical matter, which can be recognized, produced and used by our human brain alone (Li and Song 2009). It is one of the most strategically important resources that organizations have (Goh and Sandhu 2013) and may affect community development (Sakolnakorn and Naipinit 2013). Knowledge is also regarded as a source of power. It not only assumes the authority of "truth" but also processes the power to make it true once it is applied in the real world. Knowledge, entails constraint, regulation and the discipline of practice (Selvadurai et al. 2013). Where does new knowledge come from? Some researchers are of the view that growth in knowledge is exogenous to the economy. It is a result of economic progress under the lens of new growth economics. Others concede that knowledge produces knowledge. It is produced by our decisions to invest resources and produce new knowledge. In the process of knowledge creation, learning from others is unavoidable. Although information is kept private, the current theories of economic growth emphasize the role of information externalities, i.e. information produced for private gain ends up as a by-product to the social pool of information unintentionally. This in turn serves as an input in knowledge creation and goods production (Arrow 1994).

Nevertheless, its intangibility in nature, the high transaction costs in processing knowledge (Li 2014) also leads to the following concerns:

1. Some knowledge is difficult to replicate and imitate (Goh and Sandhu 2013);
2. It is transferred to the others and reproduced via sharing and studying (Li and Song 2009).
3. Knowledge sharing process is voluntary, extra effort requires on top of job requirement (Li and Poon 2009; Li and Song 2009).
4. Motivation is needed to smoothen the process of knowledge sharing (Li and Poon 2009).
5. If an organization does not possess the ability to motivate its employees to share knowledge, the knowledge inside the organization would not be optimized practically (Qayyum et al. 2011).

Thus, many organizations are looking for factors that motivate their employees to acquire continuous learning with inherent knowledge and secure their existence in an organization (Qayyum et al. 2011).

Knowledge value increases when we share our knowledge (Ah Pak and Li 2009). As we engage in knowledge sharing, organization members exchange ideas.

When the ideas collide and enlighten, organizational members convert implicit knowledge to explicit knowledge (Lei and Chen 2011). Under this process, not only those who receive it gain information, the feedback adds value to the original knowledge sharer which in turn creates an exponential growth in knowledge (Ah Pak and Li 2009).

5.3 E-Learning

E-learning has been viewed as is an emerging mission in modern educational technology (Basha and Dhavachelvan 2010). Unlike the traditional face-to-face learning process which is delivered by lecturer, e-learning process is created by interaction with digitally delivered content, supported and serviced by various Information and Communication Technology (ICT) to facilitate learning (Riad et al. 2009) at anytime and anywhere. It can also be viewed as a form of electronic exchange of information for learning aided by ICT (Hrbáčková and Vávrová 2012; Tawil et al. 2012).

The history of e-learning can be dated back to early 1960s Programmed Logic for Automatic Teaching Operations developed by the Computer-based Education Research Laboratory of the University of Illinois at Urbana-Champaign (Basha and Dhavachelvan 2010). The development of e-learning, however, was very slow after that. The development of World Wide Web, improvement in ICT and substantial drop in personal computers and fixed internet's prices in the 90s revolutionized in the idea of e-learning. As many households afford to buy a computer for leisure, study and work (Abedin 2011), many institutions nowadays use e-learning to improve learners and lecturers' performance. All these change the facet of education: various learning activities can be learnt via virtual classroom. It is predicted that e-learning has the potential to become a low cost education tool. For example, Newcastle University has implemented Mathematics and Statistics e-learning courses because of more function and control in the content, the ability to deliver e-assessments to local schools with a variety of platforms (Tawil et al. 2012). Similar actions have also taken place in University of South Australia.

The development of Web 2.0 where web browsers are allowed to read, write and edit the online content (Li and Poon 2011; Li and Ah Pak 2010; Li 2010b) has revolutionized the concept of e-learning. While the use of Web 2.0 is quite limited for formal education purposes, they have already used in the students' personal world (Abedin 2011). For example, some of them learnt econometric modelling techniques and software via YouTube. In Adelaide's construction industry, many project managers have joined the LinkedIn to receive updated information. There are also construction safety groups in LinkedIn with many construction practitioners/anyone who is interested in construction safety issues to ask questions and receive answers when they face different types of site safety problems. For example, Roberts (2014) from UK & Europe Rail BDM Iseeu Global Ltd post a question "[i]s Fatigue and Lone working a real issue within construction. We work in the health market and within the rail industry. Within rail fatigue management is a huge issue?" and successfully received three answers from safety trainer in Russia and safety expert.

5.4 M-Learning: The Post E-Learning Era

As there is wider and better mobile networks as well as cheaper mobile devices, mobile learning (m-learning), a modern form of e-learning (Binsaleh and Binsaleh 2013), delivered through mobile devices such as smart phones, Palms, pocket personal computer, Personal Digital Assistants, tablet computer, digital cell phones, and other handheld devices (Manuguerra and Petocz 2011; Binsaleh and Binsaleh 2013), offer support for effective learning and performance-based assessment (Binsaleh and Binsaleh 2013). In short, the definition of m-learning is the educational use of mobile device. Current trend witnesses the distance education throughout the world with mobile device to support the teaching and learning process (Ismail and Azizan 2012).

M-Learning is convenient and flexible. It allows distance learners to access information, learn anywhere and anytime at affordable costs with wireless technologies (Binsaleh and Binsaleh 2013). This is particularly important to the some learners who cannot learn something due to their geographical limitation. This certainly have some important implications for workers on sites. As workers always move from one site to another without fixed working locations, it is unwise to request them to stay in one place to gather for any new information. Previous research has already shown that this problem affects the effectiveness of some construction safety management implementation on site (especially the large one) such as KYT in Hong Kong (Li and Poon 2013), m-learning definitely helps overcome geographical problems as such.

On top of that, earlier studies have shown that m-learning may enhance learners' interaction, provide wider and more flexible access to education. It also enhances active, cooperative and collaborative learning environment. By strengthening the synergistic relationships from the system, it allows learners to deliver aggregate, choose and transform their own knowledge that they need. In view of the flexibility and convenience, m-learning is regarded as the future of distance education (Ismail and Azizan 2012). From this perspective, proper adoption and well-designed m-learning in construction safety education may also achieve the same goal. This is of particular importance to construction workers who need to cooperate with the others on sites and work as a team for a period of time. When they learn via m-learning, the collaborative environment in the virtual World may extend to our real World on site.

5.4.1 The Potential of M-Learning: A Glance at Our Fast Growing Market

The fast growth of new mobile devices such as tablets, mobile phones and progression in wireless technology has increased the potential of mobile learning to become an effective learning tool. Malaysian Communication and Multimedia Commission reports that the major mobile phone users range from 20 to 49 years. Digital Media shows that there are 100.8 mobile phones per 100 Malaysians which implies that some Malaysian may have more than one mobile phone. Malaysian 3G subscriptions grew tremendously from 427 in 2006 to 4,366 in 2008. In view of the increasing mobile phone owners, educators look into the possibility of integrating m-learning to university academic programs (Hussin et al. 2012).

5.4.2 Mobile Learning Worldwide

In view of the rapid development of m-learning, there is an expansion on the research studies with regards to the impact of m-learning on learners' learning process. For example, MOBIlearn project, a Worldwide European-led research and development project, explores the informal problem-based and workplace learning by various mobile technologies (Binsaleh and Binsaleh 2013). An Australia research find that learners are quite positive on m-learning with devices such as iPad despite most of them do not believe that their learning have improved (Manuguerra and Petocz 2011). An iPad initiative in a US graduate management program notes that despite learning outcomes have not significantly improved, m-learning enhance environmental sustainability, convenience and flexibility. In Canada, the use of an iPad in multiple-choice examination show that all participants like the idea as it saves paper resources. Nevertheless, some students report there is extra stress caused by the potentially unreliable or unstable internet. They are worried that their responses are recorded and submitted incorrectly (Manuguerra and Petocz 2011).

A lot of the current mobile apps on safety are mainly used for recording information. For example, Heavy Construction Systems' specialists Inc. develop Safety Field Manager for Mobile to allow the manager to record inspection information. Safety meeting Apps is developed to record safety meeting, active/inactive employers' incident on site (Safety Meeting App 2014).

5.5 Transaction Costs, M-Learning and E-Learning

"The New Institutional Economics is preoccupied with the origins, incidence, and ramifications of transaction costs" (Williamson 1979). Transaction costs, under the lens of institutional economics, are any costs that do not exist in one man economy

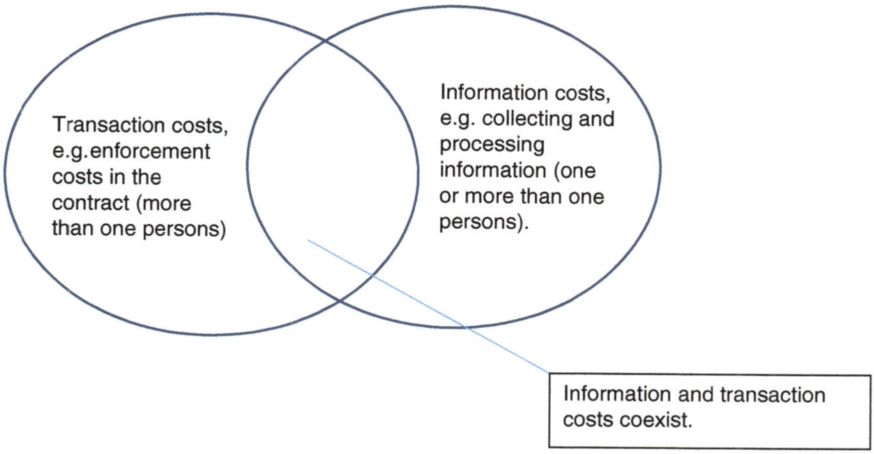

Fig. 5.1 Relationship between information costs and transaction costs (Li 2010a)

(Li 2010c). The costs of obtaining and searching for information costs are two of the examples of transaction costs. Nevertheless, there are some differences between transaction costs and information costs (Li 2010a) as information costs may exist in one man economy and transaction costs refer to costs that only exist when there are more than one persons (Fig. 5.1). From this perspective, the costs involved in safety knowledge sharing among workers or between safety trainers/safety offices and workers are classified as transaction costs.

5.6 Mobile Apps for Construction Safety

Previous search research of 249 construction firms reveal that construction businesses use mobile applications (Apps) for updating project status, and assigning tasks (22 %), scheduling (29 %), organizing business contacts (40 %) as shown. The use of mobile devices allows construction workers to instant report on sites, make decisions quickly so as to reduce time-consuming errors. It is reported that some companies even create mobile zones which gives room to workers for testing the new mobile apps for construction activities on sites (Maskara 2014). As the abovementioned research studies have shown the possibility of using mobile apps on sites, it also opens a new research agenda of whether there are some safety apps that have been designed to enhance construction safety learning, knowledge and administration (Fig. 5.2)

Canvas (2014), Simple But Needed (2014) and Fulcrum (2014) develop construction safety inspection checklist such that the new safety officers know what they should include when they perform safety inspections, lower their time in searching for the relevant information from other safety officers or from the previous safety officer who has already resigned from the construction companies/contractors. Experienced safety officers can use them as reminder to prevent leaving out any important safety inspection items. Heavy Construction Systems Specialists Inc (2013), Monarch Construction Safety App (2014a) and Safety Meeting App (2014) record information such as near misses, accidents and safety meetings. The accidents report allow workers to learn from mistakes and avoid accidents happen in the future (Monarch Construction Safety App 2014a).

Moreover, as compared to the traditional knowledge sharing methods, which include face-to-face teaching and learning, black and white notes for workers as well as safety knowledge sharing via telephone, mobile apps make sharing safety videos which can be re-shown for many times until the workers understand the whole process. Furthermore, as workers' attention time spans are short and have high time costs, 2 to 3 minutes video before using a tool which can share the safety knowledge to workers, lower the chances of accidents happen on sites (Scott development Group 2014). Video as such also lower the transaction costs, the costs of obtaining knowledge, which require double conscience of time. As both the safety trainers and workers have high time costs, mobile app can substantially lower the transaction costs of knowledge sharing.

United States department of Labor (2014), for example, has developed an App which allows construction workers and managers to calculate the heat index on site so as to access the level of risk when the workers work outdoor workers according to the heat index. HT NIOSH Lift Calculator allows workers to use the app to calculate the weight limit for a job (App Shopper 2014). Employers Mutual Casualty Co. and Affiliates (2013) develops LiftRight app, recommended weight limit and lifting index can be calculated by entering workers' lift data. By providing recommendations based on the calculations, it improves workers' safety (Employers Mutual Casualty Co. and Affiliates 2013). As compared to the traditional face-to-face recommendations about heat stress, mobile apps as such lower the costs of searching the information from a safety expert substantially (Fig. 5.2).

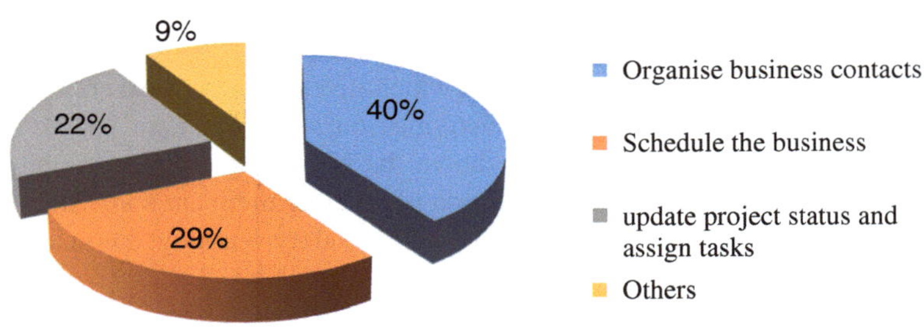

Fig. 5.2 Use of mobile apps on construction sites (Maskara 2014)

Table 5.1 Mobile apps for construction safety (author's table)

Software provider	Name of the software	Country	Uses of the construction safety apps
Mobile inspection (MSI Data 2014)	VEIL inspection software	US	It ensures compliance with safety or regulatory standards, procedures and policies
			It provides calibration or specification of equipment
			It provides equipment troubleshooting defects or failures on sites
			It monitors security or safety of equipment and personnel
Heavy Construction Systems Specialists Inc (2013)	Safety field manager for mobile	US	There are more than 500 pre-loaded tool boxes in field manager of the apps which allow the foremen on sites to record information such as near misses, inspections and meetings
			The safety field manager integrates with HeavyJob such that workers from all field of operations are reading the same page
Safety Meeting App (2014)	Safety meeting app	US	It generate PDF reports which can be printed and downloaded
			It provides date stamps for each of the safety meetings
			It covers many types of trades
			It records the disciplinary actions with electronic signatures
			It records the attendance of the employees
			It records the incidents or accidents
Canvas (2014)	Construction safety inspection report—safety-link mobile app	US	This mobile app provides construction safety inspection checklist which is used to promote and ensure safe working conditions on sites
			The checklist covers information about housekeeping, fire safety, lighting exits, documentation, permits and so on
			The app also process the capability to capture images, GPS location and a signature

(continued)

Table 5.1 (continued)

Software provider	Name of the software	Country	Uses of the construction safety apps
Canvas (2014)	Safety weekly inspection of scaffold—safety-link mobile app	US	It ensures workers' safety and health by performing weekly inspection checklist for scaffolding quickly and with ease with smartphone or tablet
			It records all the defects and arranges for competent scaffolding workers to rectify the defects
			It includes photo and signature capture tools
Simple But Needed (2014)	Mobile construction app	US	It helps construction managers to monitor construction safety on site with various mobile devices
			Inspectors can use the SBN's library checklists or generate checklists with administrator's desktop tool
			It allows inspectors to complete checklists, incorporate photos, generate reports and notify construction managers of various safety issues on site
			The safety inspectors can also be notified with regards to any safety problems at an instant via an electronic record
			Because workers can use portable mobile devices such as iPhones and other smart phone devices, it eases the entire construction crew to ensure construction safety on sites
			The app can be configured to record safety tool box meetings
			It enhances safety support for injured workers who work alone on site
Monarch Construction Safety (2014b)	Monarch construction safety app (no. 1 app)	Not specified	By assessing hazards, workers are aware of the dangers associated with tasks on sites. Accidents can be examined properly and avoided in the future
			Assessing hazards on the job site with the app with is an easy and quick way to record and share the information of the hazards

(continued)

Table 5.1 (continued)

Software provider	Name of the software	Country	Uses of the construction safety apps
Monarch Construction Safety App (2014a)	Monarch construction safety app (no. 2 app)	Not specified	Accident and near miss reports record the near-misses and accidents on sites. It helps workers learn from mistakes and prevent the tragedies from happening again. It also keeps a record of who should be reported to and how the accident was dealt with
			Tool-box meetings are used to share safety knowledge and concerns on sites. It records the workers' attendance records, informs us who is aware of what information, and who should speed up their jobs
			Safety observation sheets records workers' positive or negative aspects of their jobs on site. This helps the safety program remedy the problems before they become accidents and reinforces which program has already implemented
			Tool tag-out sheets notify which tool should be deemed out-of-service, requires repair or a previously broken tool is now safe to use again
Fulcrum (2014)	The construction safety inspection app		The construction safety inspection app provides a checklist to report hazards when we do safety audit and site visit
			This app provides construction teams and site managers a simple way to report conditions quickly and accurately when we conduct safety reviews for tracking unsafe working conditions or misuse of equipment, documenting observed hazards with photo documentation

(continued)

Table 5.1 (continued)

Software provider	Name of the software	Country	Uses of the construction safety apps
Centers for Disease Control and Prevention (2014)	NIOSH ladder safety app for mobile devices	US	The app provides a graphic-oriented guide and a multimodal indicator for ladder inspection, selection, positioning, renovating and safety use
			The app is available in English and Spanish
United States department of Labor (2014)	Heat safety tool	US	The app allows workers and supervisors to calculate the heat index on site and a risk level is shown to outdoor workers based on the heat index
			With a simple click, reminders about protective measures such as short breaks that should be taken to protect workers from heat stress related illness, drinking sufficient fluids, planning for what to do in an emergency, training on heat stress related illness signs and symptoms, monitoring workers for symptoms of heat-related illness and adjusting work operations
App Shopper (2014)	HT NIOSH Lift calculator	US	The impact of manual material handling represents the single largest contributor in construction industry. According to Liberty Mutual Workplace Index, overexertion injuries which include lifting injuries account for $13.4 billion direct costs
			Workers can use the app to calculate the weight limit for a specific task or job on sites
			It helps workers to prioritize the jobs that require ergonomic intervention
American Red Cross (2014)	First aid app	US	It provides expert advice for first aid emergencies
			It includes videos, simple step-by-step advice and interactive quizzes of the first aid

(continued)

Table 5.1 (continued)

Software provider	Name of the software	Country	Uses of the construction safety apps
iTunes (2014)	Safety Pro	US	It offers an efficient and secure method to keep information electronically for compliance audits, future reference, accident investigations and pre-shift inspection
CNet (2014)	Dropped objects calculator	Not mentioned	Enter the estimated fall distance and the weight of the object, the calculator will tell you the potential impact on humans
			The results are shown with standard safety terminology and joule impact of the fall
			Understanding the potential impact of a falling object is the key in work planning, safety awareness and accident prevention
			The calculator was developed by the website developer of Risk-Report (http://www.aboutriskreport.com) with DROPS
			RiskReport believes that a safe working environment requires current-state-of-art solutions and tools
			DROPS is an industry-wide initiative which focuses on falling objects prevention, with an ultimate goal to deliver a falling objects prevention strategy across industry
			DROPS is a global workgroup in the energy sector with more than 70 contractors, operators, service companies and industry bodies, all sharing enthusiasm and commitment for the goal of falling object prevention so as to create a safety working environment

(continued)

Table 5.1 (continued)

Software provider	Name of the software	Country	Uses of the construction safety apps
Scott Development Group (2014)	iConstructSafe	US	iConstructSafe is a collection of 2–3 min safety reminder videos which can be stored and viewed via an iTouch, iPhone or iPad in absence of wifi
			The app includes a selection of safety videos on air, hand, power, gas, tool selection and safety tips about the tool on site
			It also includes videos on PPE, falls prevention, ergonomics and so on
			The videos are designed to inform workers on how to use tools safely before they use the tools
			It features a safety tip of the day
Employers Mutual Casualty Co. and Affiliates (2013)	Liftright	Not mentioned	By entering the workers' lift data, it calculates the recommended weight limit and lifting index
			It improves workers' safety by providing recommendations based on the calculations
			It allows users to attach photos or notes about lift information such that the users can read and view again at anytime and anywhere

Note There are a lot more safety apps available in Google Play Store in the website https://play.google.com/store/search?q=construction%20safety%20app&c=apps

5.7 Conclusion

The popularity of mobile apps change the facet of education and learning. It revolutionized e-learning to m-learning. Workers can learn and share construction safety knowledge via video with the help from various forms of mobile technology such as iPhones, iPads, iTouch and tablets. It reveals that many of the construction mobile apps for safety are used to record the construction accidents and make checklists for the safety officers, share the safety knowledge via videos and calculate the various indices to estimate the safety risks at work during high temperature and lifting process. The results show in Table 5.1, however, also reveals that most of the apps are developed by apps companies in the US. It implies that there is a large room for safety apps in other places to develop the apps which are suitable for local needs.

References

Abedin, B. (2011). Web 2.0 and online learning and teaching: A preliminary benchmarking study. *Asian Social Science, 7*, 5–12.

Ah Pak, D. H., & Li, R. Y. M. (2009). Win-win knowledge management strategies between a new born baby and a giant in Academia. *The 6th International Conference on Knowledge Management Proceedings*, 2009, Hong Kong, China.

American Red Cross. (2014). *American Red Cross*. Retrieved August 31, 2014, from http://www.redcross.org/mobile-apps/first-aid-app.

App Shopper. (2014). *HT NIOSH Lift calculator*. Retrieved August 31, 2014, from http://appshopper.com/reference/ht-niosh-lift-calculator.

Arrow, K. J. (1994). Methodological individualism and social knowledge. *The American Economic Review, 84*, 1–9.

Basha, M. S. S., & Dhavachelvan, P. (2010). Web service based secure e-learning management system—EWeMS. *Journal of Convergence Information Technology, 5*, 57–69.

Binsaleh, S., & Binsaleh, M. (2013). Mobile learning: What guidelines should we produce in the context of mobile learning implementation in the conflict area of the four southernmost provinces of Thailand. *Asian Social Science, 9*, 270–281.

Canvas. (2014). *Construction, repair and improvement mobile apps and forms*. Retrieved August 31, 2014, from http://www.gocanvas.com/mobile-forms-apps/37-Construction-Repair-Improvement/show_category.

Centers for Disease Control and Prevention. (2014). *Fall injuries prevention in the workplace*. Retrieved August 31, 2014, from http://www.cdc.gov/niosh/topics/falls/.

Chang, C. C., & Hsiao, K. C. (2011). A SOA-based e-Learning system for teaching fundamental information management courses. *Journal of Convergence Information Technology, 6*, 298–305.

CNET. (2014). *Dropped objects calculator*. Retrieved August 31, 2014, from http://download.cnet.com/Dropped-Objects-Calculator/3000-2094_4-75277066.html.

Dutta, R., Kundu, A., Dattagupta, R., & Mukhopadhyay, D. (2009). An approach to web page prediction using Markov model and web page ranking. *Journal of Convergence Information Technology, 4*, 61–67.

Employers Mutual Casualty Co., and Affiliates. (2013). *LiftRight*. Retrieved September 7, 2014, from https://play.google.com/store/apps/details?id=com.emcins.LiftRight.

Fulcrum. (2014). *Construction safety inspection*. Retrieved August 31, 2014, from http://fulcrumapp.com/apps/construction-safety-inspection/.

Goh, S. K., & Sandhu, M. S. (2013). Affiliation, reciprocal relationships and peer pressure in knowledge sharing in public Universities in Malaysia. *Asian Social Science, 9*, 290–298.

Heavy Construction Systems Specialists Inc. (2013). *What is HCSS mobile apps?* Retrieved July 31, 2014, from http://www.hcss.com/products/construction-apps/overview.

Hussin, S., Manap, M. R., Amir, Z., & Krish, P. (2012). Mobile learning readiness among Malaysian students at higher learning institutes. *Asian Social Science, 8*, 276–283.

Hrbáčková, K., & Vávrová, S. (2012). Subjective conception of students' self-regulated learning from the perspective of a beginner teacher. *Asian Social Science, 8*, 228–239.

Ismail, I., & Azizan, S. N. (2012). Distance learners' needs on interactivity in sms-based learning system. *Asian Social Science, 8*, 119–128.

ITunes. (2014). *Safety Pro*. Retrieved August 31, 2014, from https://itunes.apple.com/us/app/safety-pro/id465612643?mt=8.

Jansen, B. J., Spink, A., & Saracevic, T. (2000). Real life, real users, and real needs: a study and analysis of user queries on the web. *Information Processing and Management, 36*, 207–227.

Lei, J., & Chen, J. (2011). How Chinese enterprises effectively establish self-management team. *Asian Social Science, 7*, 136–140.

Li, B., & Song, W. (2009). A discussion on the relationship and mutual effects of knowledge management and information management. *Asian Social Science, 5*, 166–171.

Li, R. Y. M. (2010a). *Factors influencing developers' decision to sell housing units with fittings: Empirical evidence from China*. Hong Kong: The University of Hong Kong.

Li, R. Y. M. (2010b). Knowledge sharing by web 2.0 in real estate and construction discipline. In M. Anandarajan & A. Anandarajan (Eds.), *E-research collaboration: frameworks, tools and techniques*. Heidelberg: Springer.

Li, R. Y. M. (2010c). A study on the impact of culture, economic, history and legal systems which affect the provisions of fittings by residential developers in Boston, Hong Kong and Nanjing. *International Journal of Global Business and Management Research, 1*, 131–141.

Li, R. Y. M. (2012). Construction accidents compensation and game theory analysis on mobile safety knowledge sharing among generation Y in Hong Kong. *Romanian Review of Social Sciences, 3*, 3–12.

Li, R. Y. M. (2014). *Law, economics and finance of the real estate market—A perspective of Hong Kong and Singapore*. Heidelberg: Springer.

LI, R. Y. M., & Ah Pak, D. H. (2010). Resistance and motivation to share sustainable development knowledge by web 2.0. *Journal of Information and Knowledge Management, 9*, 251–262.

Li, R. Y. M., & Poon, S. W. (2009). Future motivation in construction safety knowledge sharing by means of information technology in Hong Kong. *Journal of Applied Economic Sciences, 4*, 457–472.

Li, R. Y. M., & Poon, S. W. (2011). Using web 2.0 to share the knowledge of construction safety as a public good in nature among researchers: The fable of economic animals. *Economic Affairs, 31*, 73–79.

Li, R. Y. M., & Poon, S. W. (2013). *Construction safety*. Heidelberg: Springer.

Manuguerra, M., & Petocz, P. (2011). Promoting student engagement by integrating new technology into tertiary education: The role of the iPad. *Asian Social Science, 7*, 61–65.

Maskara, N. (2014). *Top 5 construction safety apps*. Retrieved August 31, 2014, from http://www.mysafetysign.com/blog/top-5-construction-safety-apps/.

Monarch Construction Safety App. (2014a). *The no. 2 app*. Retrieved August 31, 2014, from http://monarchsafetyapp.com/no2.htm.

Monarch Construction Safety App. (2014b). *The no. 1 app*. Retrieved August 31, 2014, from http://monarchsafetyapp.com/no1.htm.

MSI Data. (2014). *Mobile construction apps*. Retrieved July 31, 2014, from http://www.msidata.com/mobile-construction-software.

Norman, H., Din, R., Nordin, N., & Ryberg, T. (2014). A review on the use and perceived effects of mobile blogs on learning in higher educational settings. *Asian Social Science, 10*, 209–222.

O'cass, A., & Fenech, T. (2003). Webretailing adoption: Exploring the nature of internet users webretailing behaviour. *Journal of Retailing and Consumer Services, 10*, 81–94.

Qayyum, A., Sukirno, M., & Mahmood, A. (2011). A preliminary investigation of employee motivation in Pakistan's banking sector. *Research and Practice in Human Resource Management, 19*, 38–52.

Riad, A. M., El-Minir, H. K., & El-Ghareeb, H. A. (2009). Review of e-learning systems convergence from traditional systems to services based adaptive and intelligent systems. *Journal of Convergence Information Technology, 4*, 108–131.

Roberts, M. (2014). *Is fatigue and lone working a real issue within construction. We work in the health market and within the rail industry. Within rail fatigue management is a huge issue*. Retrieved September 7, 2014, from https://www.linkedin.com/groupItem?midToken=AQEnbCfqde5Dqw&type=member&fromEmail=fromEmail&trk=eml-b2_anet_digest-group_discussions-18-grouppost-disc-1&ut=3FrvDDCHl3C6o1&item=5913606861730381824&gid=137591&view=.

Safety Meeting App. (2014). *Safety meeting app*. Retrieved July 31, 2014, from http://www.safetymeetingapp.com/.

Sakolnakorn, T. P. N., & Naipinit, A. (2013). Guidelines for the management of community enterprises in the Songkhla lake basin of Thailand. *Asian Social Science, 9*, 166–173.

Samah, B. A., Shaffril, H. A. M., D' Silva, J. L., & Hassan, M. A. (2010). Information communication technology, village development and security committee and village vision movement: A recipe for rural success in Malaysia. *Asian Social Science*, *6*, 136–144.

Scott Development Group. (2014). *iConstructSafe*. Retrieved from https://itunes.apple.com/us/app/iconstructsafe/id549574287?mt=8&ign-mpt=uo%3D4.

Selvadurai, S., Choy, E. A., Maros, M., & Abdullah, K. (2013). Shifting discourses in social sciences: Nexus of knowledge and power. *Asian Social Science, 9*, 97–106.

Simple But Needed. (2014). Web based construction management software.

Tawil, N. M., Ismail, N. A., Asshaari, I., Osman, H., Nopiah, Z. M., & Zaharim, A. (2012). Learning process in mathematics and statistics courses towards engineering students: e-learning or traditional method? *Asian Social Science, 8*, 128–133.

United States Department of Labor. (2014). *Heat safety tool*. Retrieved August 31, 2014, from https://www.osha.gov/SLTC/heatillness/heat_index/heat_app.html.

Williamson, O. E. (1979). Transaction-cost economics: The governance of contractual relations. *Journal of Law and Economics, 22*, 233–261.

Chapter 6
Supply and Demand of Construction Safety Regulations in Developing Countries

Rita Yi Man Li, Jian Zuo and Terdoo Fanen

Abstract Construction industry plays an important role in developing countries' economic development. It is often viewed as one of the major economic engines in economic growth. A lot of these countries, however, are efficient-driven which often undermine the safety importance on site. This chapter firstly reviews the economic role of construction industry in developing countries, followed by an analysis on safety regulations and related authorities in China, Nigeria, Kuwait and Ghana. It is found that 1) these countries have implemented some safety rules and regulations, 2) many of them are poorly design and 3) the enforcement is quite loose.

Keywords Construction safety regulations · Demand and supply · China · Nigeria · Kuwait and Ghana

6.1 Introduction

Previous research suggests that developing countries are efficiency-driven economies while the developed ones are innovation-driven economies. In efficiency-driven economies, institutions support industrialization to pursuit economies of scale and higher productivity. As the economy develops, the importance of industrial activity shifts towards service sector (Lakovleva et al. 2011). Therefore, construction sector as

R.Y.M. Li (✉)
Department of Economics and Finance / Sustainable Real Estate Research Center, Hong Kong Shue Yan University, Hong Kong, China
e-mail: ritarec1@yahoo.com.hk

R.Y.M. Li · J. Zuo
School of Natural and Built Environments, University of South Australia, Adelaide, Australia

T. Fanen
Federal University Dutsin-Ma, Dutsin-Ma, Katsina State, Nigeria

T. Fanen
University of Reading, Reading, United Kingdom

R.Y.M. Li, *Construction Safety and Waste Management*, Risk Engineering, DOI 10.1007/978-3-319-12430-8_6

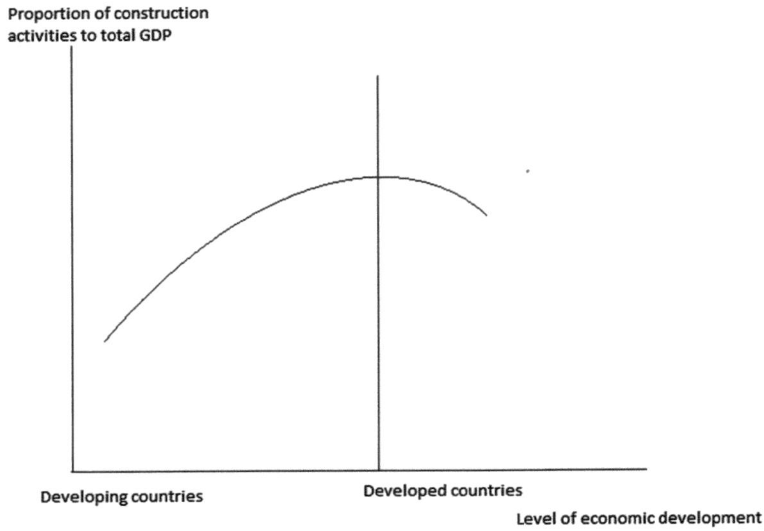

Fig. 6.1 The relationship between economic development and construction activities (author's figure)

one of the industrial activities, firstly grows in the less developed country, reaches the peak in newly industrialized country and finally decreases in advanced industrialization (Li and Poon 2013) (Fig. 6.1). "As the country is in the early stage of economic development, there is a lack of essential infrastructure such as road, railway and buildings. Everything have to be built on scratch. Nevertheless, once they are built, they can be used for many years, London Underground and Paris Metro are some of the vivid examples (Table 6.1). Although the earliest railway lines were built in the late 19th and early 20th century, they are still used as a major means of transportation nowadays. Nevertheless, if we have a quick glance at the Guangzhou Metro in China and Thailand MRT in Table 6.1, a lot of the lines there were built in recent few years or are still under construction. All these evidence the reason why construction industry plays a more important role in developing than developed countries."

Construction industry plays an important role in labour markers among the developing countries as it hires a large workforce. For instance, the industry povides 4% social employment and shares 5–6% of GDP (Li and Poon 2013). In spite of huge economic contribution, previous research shows that developing countries are often lack of effective mechanisms to implement and enforce employment laws. Many construction companies are hired on casual and temporary basis. Therefore, employment conditions are not defined properly and hence provide little protection on workers' safety and health (Kheni et al. 2006). Protection of labors' right on construction safety is little. As Kheni et al. (2008) comments: "The institutional and legal governance frameworks on occupational health and safety in developing countries have little impact… The majority of

Table 6.1 Years of underground railway construction in London, Guangzhou and Thailand (Li 2013)

London metro		Paris métro			Guangzhou metro		Thailand MRT	
Name	First operated		Opened	Last extension	Name of the line	Year of opening		Year of opening
Bakerloo line	1906	Line 1	1900	1992	Guangzhou east railway station	1997	Chaloem Ratchamongk-hon line	2004
Central line	1900	Line 2	1900	1903	Jiahewanggang	2002	Purple line	2014
Circle line	1871	Line 3	1904	1971	Airport South/Tianhe coach terminal	2005	Pink line	2014
District line	1868	Line 3bis	1971	1971	Huangcun	2005	Blue line extension	2015
Hammersmith & City line	1864	Line 4	1908	2013	Jiaokou	2009	Orange line	2019
Jubilee line	1979	Line 5	1906	1985	Fenghuang Xincun	2003	Brown line	2019
Metropolitan line	1863	Line 6	1909	1942	Kuiqi Lu	2010	Grey line	2019
Northern line	1890	Line 7	1910	1987	Linhexi	2010	Yellow line	2022
Piccadilly line	1906	Line 7bis	1967	1967			Light blue line	2029
Victoria line	1968	Line 8	1913	2011				
Waterloo and City line	1898	Line 9	1922	1937				
		Line 10	1923	1981				
		Line 11	1935	1937				
		Line 12	1910	2012				
		Line 13	1911	2008				
		Line 14	1998	2007				

contractors are SMEs operating within their domestic markets where enforcement of health and safety standards and labour standards is very lax. Enforcement of health and safety regulations remains a problem due to lack of adequate resources available to government institutions responsible for occupational and safety administration…the limited resources available to enforcement agencies and prevention services as the main factor contributing to poor health and safety conditions…Also, there remains an acute need for contract provisions to support the enforcement of labour laws in developing countries. These problems are compounded by shortages of skilled labour."

6.2 Demand of Construction Safety Measures in Developing World

Construction safety is often a persistent problem in developing countries. Previous research results showed that about 1.2 million people died of accidents at work and diseases every year and 100 times more people were ill or injured because of work (Li and Poon 2013). Some of the construction projects are of higher safety risk in nature, for instance, dam building, tunnel boring and demolition may lead to higher probability of death and injury. Thus, construction workers have higher probability of dying at work or having a major injury than any other industries (Snashall 1990).

Contractors in developing countries, in particular, the indigenous ones face many constraints which hamper the effective safety and health management. They may suffer from unregulated practices on construction sites, poor infrastructure, traditional methods of work, traditional methods of construction, corruption, communication problems due to low literacy level, absence of equipment, poor site security, extreme weather conditions and improper use of equipment. Moreover, as many workers are illiterate in developing countries, there are difficulties in training, leading to another barrier to effective safety and health management (Kheni et al. 2006). For example, the industry accounts for 42 % of all the occupational fatalities in Kuwait (Chua and Goh 2004). In Nigeria, construction industry's fatality rate reaches 20.1 % (Hamalainen et al. 2006). In Kuwait, construction industry is the most hazardous industry (Al-Humaidi and Tan 2010). In Ghana, the industry is one of the industry with the highest accident rate even though the number of employees are not the highest (Kheni et al. 2006). It is reported that 56 out of 1,120 construction accidents were fatal which was higher than International Labour Organization (ILO)'s estimation for developing countries (Li and Poon 2013) and the European Union. The European Union accident statistics is only 1.5 fatalities per 1,000 accidents. Despite many accidents in construction sector have not yet reported (Kheni et al. 2008). Likewise, report on workers' compensation which shows that the sector had poorer health and safety performance record than most of the other industries (Kheni et al. 2006).

Many of the previous research criticize the poor safety regulation in developing countries and cities is the major causes of higher accident rates than the developed countries. Some of them reveal that there is a lack of relevant regulation or there is no enforcement at all. Given the relative low compensation for injured or even death of the workers and efficiency-driven societies, many construction companies are lacking of incentive to provide adequate safety measures, construction sector remains one of the most dangerous work in the developing World (Li and Poon 2013) and the economic costs of construction accidents are very high. It occurs due to the loss of experienced and trained workers; higher insurance cost after accidents, disruptions to work progress which may lead to an extension of time. Hence, an economic incentive exist to prevent construction accidents (Choudhry, Fang, & Mohamed, 2007) with the help of various safety measures.

6.3 Supply of Construction Safety Regulations in Developing Countries

In the following section, we study four four developing countries' safety regulations: Nigeria, China, Ghana and Kuwait. Among all of them China was the country with highest GDP in 2010. Nevertheless, Kuwait stood the highest in GDP per capita as it has a relative low population. As Kuwait is recovering from the destruction of war and economic activities are limited in oil industries, it remains in the list of developing countries. In Ghana, a thriving democracy, commitment to good governance and development of the private sector powers a rapid economic growth over the past 10 years. As a matter of fact, it is one of the fastest growing economies in sub-Saharan Africa (Table 6.2). Construction contributes up to 8.8 % of GDP and employs 1.4 % of the country's work force with a significant proportion of formal employment (Kheni et al. 2010). In the following section, we aim to review the safety regulations, issues and solutions of the four selected developing countries.

Table 6.2 Background of the five selected developing countries (TheGlobalEconomy.com 2013)

	GDP (billions of 2000 US dollar in 2013)	GDP per capita, purchasing power parity in 2012
Nigeria	190.6	5439.6
China	4864.0	10756.5
Ghana	19.8	3668.0
Kuwait	96.6[a]	84187.7

[a] The data of Nigeria refers to year 2012

6.3.1 Nigeria

6.3.1.1 Role of the Construction Industry

According to the Nigerian Engineering Leadership Forum (NELF), Nigeria's construction industry is one of the largest and fastest growing industries in Africa; and has significantly contributed to Nigeria's gross domestic product (GDP) growth over the decades. It is also the fastest growing sector of the Nigerian economy next to the leading telecommunications, with an annual growth rate of 13 % in 2012. In the next decades, it is expected that the Nigerian construction industry will be the fastest growing in the world. Presently, the value of construction as a percentage of government capital expenditure in Nigeria is approximately 80 % (The Nigerian Engineering Leadership Forum 2013). Therefore, as the Nigerian economy is fast growing and presently assuming the leading position in the African continent, and as the income and wages of civil servants and the political class continue to rise, there are possibilities that the demands for construction projects generally will rise significantly thus paving way for a huge demand for and supply of construction safety regulations in the construction sector.

6.3.1.2 Related Laws and Regulations

Despite these recent developments in the construction industry in Nigeria, the demand for and supply of construction safety measures is yet to be given the serious attention it deserves by the government and the stakeholders (Dodo 2014).This is obvious, firstly, in the previous laws and decrees (acts) enacted to regulate and enforce compliance with safety measures, and secondly, in the altitude of the stakeholders driving the operations of the industry (Olutuase 2014). The Factories Act of 1958 referred to as the colonial legislation, the Workman's Compensation Act of 1987, the Labour Act of 1990, the Workman's Compensation Act of 2004, the Employee's Compensation Act of 2011 were all genuine efforts to ensure proper regulation and enforcement of safety measures in the construction industry. However, these efforts were not sufficient and inadequate to regulate and enforce compliance in the industry. This is because in some of these laws/acts (e.g. the Factories Act of 1987 and others), construction industry is completely excluded in the definition of its premises (Idubor and Oisamoje 2013; Olutuase 2014); hence, the lack of regulation and enforcement of safety measures by the stakeholders as the penalties and punishments stipulated for violation of occupational safety and health (OSH) laws in Nigeria are not severe enough and insufficient to deterred offenders and violators of safety measures in the construction industry (Idubor and Oisamoje 2013; Kalejaiye 2013; Umeokafor et al. 2014).

The enactment of the new Bill, namely: The Labour, Safety, Health and Welfare Bill of 2012 did not only address the issue of exclusion of the construction industry in the definition of its premises, but also stipulated adequate and sufficient penalties

Table 6.3 Highlights of the key issues/challenges to of OSH regulations enforcement and ways to improve OSH in Nigeria (Umeokafor et al. 2014)

Major issues/challenges of OSH regulations enforcement in Nigeria	Solutions to improve OSH enforcement in Nigeria
Lack of skilled personnel	Recruitment and training of legal officers by the legal department
Politics	
Severity of penalties	Self-regulatory style of legal organisations should be used
The legal system	Enforcement of OSH regulations at local level
Corruption	
Lack of funding	Provide sufficient OSH information
Poor legislation	Development and adoption of updated regulations
Falling short of commitment from government	Code of practice
Insecurity	Updating OSH regulations
Insufficient information	
Technological and economic growth	
Culture	

for violation of safety measures by stakeholders in the construction industry hitherto neglected in the previous laws (Umeokafor et al. 2014). In addition, the new Bill also empowers the Federal Ministry of Labour and Productivity and the National Council for OSH to regulate, enforces and implements relevant measures in the workplace; protects properties and lives; enhances OSH awareness; carries out inspection of the workplaces and monitors the compliance of all the regulations among other OSH measures stated in the Bill (Idubor and Oisamoje 2013; The National Occupational Safety and Health Information Centre 2006; Umeokafor et al. 2014). This recent legislative development has increased the awareness and strengthened the demand for and supply of safety measures as well as regulation and enforcement of OSH laws in the construction industry in Nigeria. However this is not without challenges. For example, Umeokafor et al. (2014), highlighted some of the major challenges hampering the efficient implementation and enforcement of OHS laws in Nigerian construction industry to include these among others (Table 6.3):

6.3.2 China

6.3.2.1 Role of the Construction Industry

The construction industry is one of the largest industries in China and is a vital ingredient for the development of public infrastructure. In 2011–2012, the total value of production brought about by the construction industry reached RMB

13721.8 billion, contributing 7.0 % to the Gross Domestic Product (GDP). This was an increase of 15.1 % over the previous year's RMB 11645.3 billion. By the end of 2012 there were 75,280 Construction enterprises in China. These enterprises employed some 42.7 million people and generated RMB 13721.8 billion in gross output and RMB 2658.3 billion in added value. Tax contribution of the construction industry reached 439 billion RMB in 2012 (National Bureau of Statistics of China 2013).

In the last decade, economic growth of China has continued at a tremendous pace with an average annual growth rate of some 10 %. China's entry into the World Trade Organisation (WTO) and its successful bid to host the 2008 Olympic Games have played a major role in this double digit growth rate. Similarly, the Chinese Government's efforts to alleviate the impacts of the global financial crisis play a crucial role to the constant growth of the construction industry. The total amount of fund provided by the Government reached 4,000 billion RMB of which 45 % were injected to the infrastructure development, e.g. railway, highway and power grid (Zuo et al. 2011).

Therefore, it is imperative to focus on the safety performance of the construction industry which considers the scale of construction activities, number of people working in the industry, and the risks associated with construction projects.

6.3.2.2 Related Laws and Regulations

Laws related to safety in the construction industry include Construction Law, and Safe Production Law. According to these two laws, the Regulations on the Work Safety of Construction Projects was implemented in February 2004 to provide guidelines for construction safety practices. This Administrative Regulations clearly defined the roles and responsibilities of each stakeholder in ensuring the construction safety such as contractors, design institutes, supervision engineer, and governments. The emergency procedures are also specified. Similarly, it stated the Prevention oriented approach as the fundamental principle of safety management in construction projects. The human resource issues are also covered in this Administrative Regulations, e.g. qualification of firms and individuals. It is interesting to note this Administrative Regulations is not amended at all during the last decade (The Central People's Government of the People's Republic of China 2005).

6.3.2.3 Related Authorities

The main authority related to the construction safety in China is the Ministry of Housing and Urban-Rural Development (MoHURD) at the central government level. Under the leadership of MoHURD, Bureau of Construction Quality and Safety Supervision is established at each province, municipality and autonomous administrative region to manage construction safety related issues. Similarly, there are other authorities responsible for construction projects in specific sectors. For example, there is a specific division under the Ministry of Transport dealing with safety issues associated with transport related projects such as highway, water diversion, tunnel, and bridges. There are overlapping issues of duties amongst these authorities. The coordination between the large number of related authorities is critical for the improvement of safety performance in construction projects in China.

6.3.2.4 Key Issues and Solutions

Cheng et al. (2004) highlighted the issues related to construction safety in China as well as corresponding solutions (Table 6.4). It is worth noting that majority of these highlighted issues and corresponding solutions are management oriented rather than technology related.

Table 6.4 Issues and solutions for safety management in construction projects in China (Cheng et al. 2004)

Issues	Solutions
Shortage of contractor's budget	Increase in investment resources
Lack of attention to safety protection by workers	Increase in safety training and education for workers
Insufficient involvement of local government	Increase in reprimand against violation of safety
	Improvement in safety operations
Too complicated construction project	Increase in safety inspection
Excessively inclement weather	Direct safety management by main contractor
Low level of construction technology	
Insufficient safe operations	Increase in safety technology
Lack of attention to safety management by main contractors/project managers	Improvement in self-safety equipment
Tiredness of workers	Implementation of safety policies and programmes
Mobility of workers	No alcohol at work
Too many subcontractors	Selection of safe contractors and subcontractors
Poor quality of construction materials and equipment	Analysis of safety investment and effectiveness
Insufficient safety training	
Inadequate setting of safety level	Senior management involvement in safety management practices
	Safety profile (record) for each operative
	Enhancement of equipment management
	Implementation of accidental insurance scheme

The safety performance of construction projects is poor compared to the international standard (Tam et al. 2004). Attitude of stakeholders towards safety is one of most important factors to be considered for the improvement of safety performance in construction projects (Aksorn and Hadikusumo 2008). For instance, the attitudes of workers may affect their behaviour on site and consequently poses different levels of risks to themselves, e.g. likelihood of occurrence of accidents. The safety oriented attitude can be developed via regular trainings. Similarly, a safety management system at the international standard such as the OHSAS 18001 helps to streamline the safety management practices as well as developing safety attitudes within related stakeholders (Zeng et al. 2008).

According to Liao et al. (2014), communication plays a crucial role in maintaining a safety climate in Chinese construction projects as many workers and managers are lack of understanding of safety. Tam et al. (2004)'s findings suggest that the safety manual is not communicated well with workers. As a result, the safety management system defined at the corporate level has not been communicated well to site level. Hence, a well-defined communication mechanism is required to bridge the gap.

6.3.3 Ghana

6.3.3.1 Role of the Construction Industry

In Ghana, the construction industry contributes to 3.1 % of the total GDP. The industry is dominated by small local companies with few of them hire more than 250 employees. Nearly all the local construction companies in Ghana are run by owners, spouses, their children and other close relatives as family businesses. Large firms which hire more than 250 are mainly foreign firms and are very rare. Similar to many other developing countries, Ghana's construction's construction safety performance is poor. It has higher accident rate than many other industry (Kheni et al. 2006).

6.3.3.2 Related Laws and Regulations

Legislation on health and safety (including ILO conventions) are approved by the parliament Safety and health clauses in contract conditions form part of the legal framework of the Ghanaian occupational safety and health management system. Statutes, common law, ILO Conventions provide the basis to enforce safety and health measures on sites. The statutes which manage construction health and safety include:

1. The Factories, Office and Shops Act 1970;
2. The Labour Act 2003;
3. The Workmen's Compensation Act 1987 (Kheni et al. 2006).

6.3.3.3 Related Authorities

There are five ministries responsible for OHS in Ghana: Occupational Health Service Unit, the Environmental Protection Agency, the Labour Department, the Factory Inspectorate Department and the Attorney General's Department (Li and Poon 2013). Among them, the Labour Department and Factory Inspectorate Department are the two major government departments responsible for safety, health and workers' welfare. They are under the Ministry of Manpower Development and Employment. The Factory Inspectorate Department has the right to enforce workplace safety and health standards. It helps to ensure that all the workplaces attain the minimum requirements of safety and health prescribed by the Factories, Offices and Shops Act (Act number 328). Apart from site inspections, the department proactively promotes workplace health and safety through seminars and workshops, provides occupational safety and health information, conducts workplace surveys and registers factories (including construction sites). The labour department administers the Workmen's Compensation Act (1987) and the Labour Act (2003). Nevertheless, these two departments face logistical and human resources constraints which have adversely affect their operational efficiency. The Chief Factory Inspector find that the salary and fridge benefits are not attractive and some staff leave after basic inspection training and work for a year. The turnover rate is very high which is up to 50 % (Kheni et al. 2006).

6.3.3.4 Key Safety Issues and Solutions

It is observed that safety regulations are managed poorly by government departments (Kheni et al. 2006) and compliance of safety legislation is low. Problems of compliance with legislation are evident from the brief safety provisions in contract clauses The Chief Factory Inspector indicated the difficulty of monitoring construction safety "[u]nfortunately, construction…is one of the hazardous areas you can get workers in, but before you even get to a construction site, they have already started the work either out of ignorance of the law or their refusal to comply with the law. It is not like a factory where you know the location, and it is there for several years. If they close down they will come and tell you whereas most contractors will not register their sites and by the time the Factory Inspectorate is aware, it is near completion. Often, we don't get to construction sites to inspect the site layout and other safety and health aspects before they start construction" (Kheni et al. 2006).

Table 6.5 Issues and solutions for safety management in construction projects in Ghana (Kheni et al. 2006; Li and Poon 2013)

Issues	Solutions
The safety regulations are scattered in many different rules and regulations	One separate sections for occupation/construction safety may help
There is a lack of careful planning in OHS legal development, leading to inconsistencies, overlapping or fragmentation in OHS laws	Careful planning of the rules and regulations are needed
The compliance rate of safety legislation is low	Enforcement has to be tightened

The safety regulations and laws in Ghana are scattered in Environmental Protection Agency Act, the Mining Regulations and Labour Act. There is a lack of careful planning in OHS statutes development, leading to fragmentation, overlapping and inconsistencies in OHS laws. It also fails to defines the employees, employers and government's responsibilities clearly. Secondly, many illiterate workers have not been covered by insurance. Thirdly, many construction firms rely on temporary workers who do not belong to any form of labor unions (Li and Poon 2013) (Table 6.5).

6.3.4 *Kuwait*

6.3.4.1 Role of the Construction Industry

Construction workers make up about 9.9 % of the total labor force from 2001 to 2007. Figure 6.2 illustrates the trend for the number of total labor force and the number of construction workers in Kuwait. In spite of the relatively low number of labor force as compared to the entire industry, the percentage of accidents in the construction industry is a lot higher than all the other industries (Al-Humaidi and Tan 2010).

6.3.4.2 Related Laws and Regulations

Kuwait labour law covers labors' right which includes safety practices on sites (Al-Humaidi and Tan 2010). It states the right to join the trade union (Al-Humaidi and Tan 2010). In June, July and August, workers are prohibited to work from 12:00 to 16:00 to prevent workers from dehydration due to Summer's hot weather (Li and Poon 2013).

Construction injuries are classified into permanent disabilities, deaths and injuries. Workers are compensated according to the level of injuried severity:

1. Permanent disable workers are compensated according to the level of disability. The maximum compensation is around Kuwaiti dinar 13,333 (US$48,000) or income of 2000 days whichever is greater.

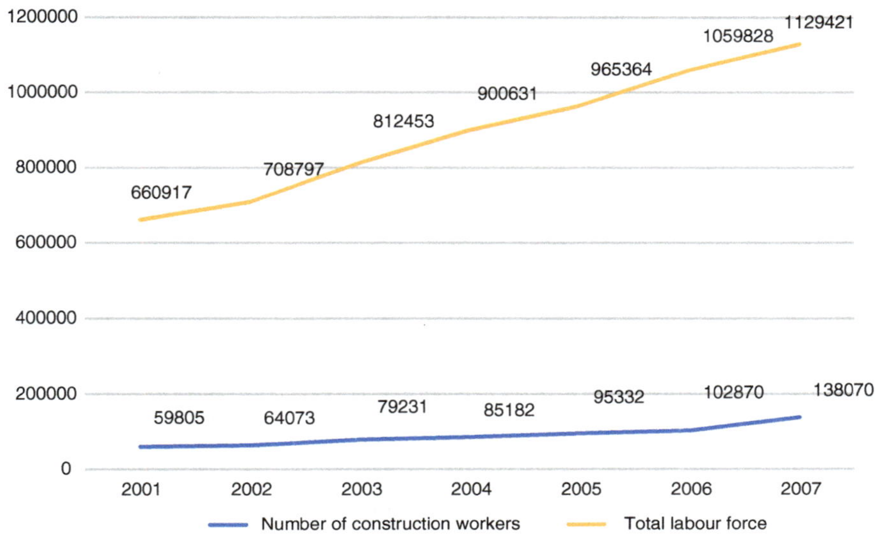

Fig. 6.2 Number of total labor force and construction labor force in Kuwait (Al-Humaidi and Tan 2010)

2. The deceased family members are compensated a salary of 1,500 days or KD 10,000 (US$36,000) (Li and Poon 2013).

Furthermore, even though Article 95 Chapter 15 of the general provisions in Kuwait private sector labor law states that government officials have to carry out site inspection, execute and implement labor safety rights, most of the safety inspections are carried out by site engineers who work for the owners or their representatives.

Firstly, rules with regards to compensation of injured or deceased workers in accidents and a competent safety officers are needed. Finally, a database which records all the nature, causes and results of the construction accidents are necessary. All these measures are used to improve on-site safety and reduce the likelihood of accidents in the future (Li and Poon 2013).

6.3.4.3 Related Authorities

The Safety Department provides safety information with regards to the proposed activity or job. It also sends a safety representative to conduct site visits to ensure temporary site access, storage and services' safety. They also provide safety posters with major instructions given to the contractor to be hung on sites on top of accident prevention methods for each of the activities's safety procedures. The Municipality charges a certain fee as an insurance for safety and work completion which is returned to the contractor when the project is completed together with a clearance certificate (Kartam et al. 2000).

KM hires safety engineers and inspectors to conduct daily site inspections on construction sites in Kuwait. These inspections include safety procedures and equipment provided by the company alongside proper disposal of construction materials and site cleaning. A safety checklist contains the major possible safety violations which aims KM safety inspectors when they do inspection. This checklist contains 29 items and is divided into three major sections:

1. Storage permits;
2. Safety procedures on construction sites;
3. Road and traffic permits (Kartam et al. 2000).

In Kuwait, when an accident happens, it is reported to the police first. The Safety Department KM deals with the construction accidents directly after it is informed by the police. After a notification from the police is received, the police will send a safety engineer to visit the site, conduct investigation (including the safety procedures), interview the witnesses and the injured victims if possible and finally write the report about the site conditions and speculate the cause of the accident. Then, the KM sends the report to police and insurance company. In general, companies violate the safety requirement are penalized by a penalty ticket or a warning ticket according to the type of violation. The penalties for each of the violation range from a few KD to a few thousand (1 KD = US$3.27) (Kartam et al. 2000).

6.3.4.4 Major Safety Issues and Solutions

Similar to many of the developing countries, they are imperfect in construction safety regulations. For example, the right to join the labor union is confined to Kuwaiti citizens only (Chap. XIII, Kuwait Private Sector, Kuwait Labor Law) despite 52 % of workers on sites are non-Kuwaitis (Al-Humaidi and Tan 2010), leading to insufficient labor protection and the labors are difficult to lodge complaints with regards to safety issues. Although thousands of safety violations have been recorded on construction sites and thousands of safety warnings are issued by KM, there are about 100 safety tickets given each year only. KM is notified late after the accident sometimes, leading to poor or incomplete safety records. One of the large companies had 30 claims over the past 3 years, however, it only paid a small amount of money as labour compensation. All these evidence the loose enforcement in construction safety (Kartam et al. 2000). The absence of unified safety regulations in the three major government departments KPW, KOC and KM adversely affects the safety enforcement on site:

1. The MPW has a safety chapter with regards to construction practices.
2. KOC has its own manual.
3. There is no standard safety manual in KM (Kartam et al. 2000).

Projects constructed by the US companies are ruled by the US, OSHA or Army Corps of Engineers safety manual (Kartam et al. 2000). They may follow the British Safety Standards when the projects are constructed by the UK companies.

Table 6.6 Key issues and solutions of Kuwait's construction safety (Kartam et al. 2000; Li and Poon 2013)

Key issues	Solutions
The right to join the trade union is restricted to Kuwaiti citizens even though 52 % of the citizens	The right to join the trade union should include non-Kuwaiti citizens
There is a lack of unified requirements and codes for local Kuwait construction safety issues	Unified codes should be provided and implemented on sites
There is a lack of detailed reports of accidents	Detail accident reports should be provided
There is a lack of qualified safety inspectors	The authority should train more safety inspectors and salary should be increased so as to retain the qualified ones
There is a loose legal enforcement on construction safety issues	Legal enforcement should be tightened

Alternatively, the may follow the Kuwait's local standard (Li and Poon 2013). Nevertheless, international standards may not necessarily suitable to the Kuwaiti work environment as the advanced industrialize countries' methods of practice may be different from those which are used in Kuwait. Having said that, following the national standards may not be a wise choice either. Many of these standards have not been updated to meet the needs of the new construction methods and technology. For instance, prohibition of wooden scaffolding has not been included in these standards. Current governmental safety inspection programs are ineffective as there is a lack of qualified inspectors (Kartam et al. 2000). It is in no doubt that the lack of up-to-date requirements and unified codes leads to construction safety problems. Furthermore, the Worker's Affair Department provides the right to the government to impose stricter construction rules, legislations and standards but fails to provide detailed investigation reports of accidents. Hence, senior managerial staff and workers cannot learn from mistakes easily (Li and Poon 2013) (Table 6.6).

6.4 Conclusion

In developing countries, construction industry is very important in economic development as many of the infrastructure such as roads, bridges and rails have not been constructed. Nevertheless, many of these countries' safety regulations are poorly designed or/and enforced. Accident compensation is also limited. For example, he there is lack of safety requirements and codes in Kuwait, piecemeal development in Ghana's safety regulations, main contractor and project managers do not pay attention on safety issues. In view of high accident rates, poor construction accidents records, huge costs of accidents which delay of construction activities imply that there are strong demand for better safety regulations. In summary, the supply of effective safety legal rules and regulations in many of the developing countries fail to meet the demand.

Acknowledgments This book chapter is a revised and substantial extension of the first author's keynote conference paper: Li, Rita Yi Man (2014) Is there any consensus on the causes of construction accidents? A critical review on digital information in academic journal database, the Fifth International Conference on the Applications of the Digital Information and Web Technologies, 17–19 February 2014, Tanjore and Chennai, India. The author would like to thank the conference organizer for granting the right to publish this chapter with substantial revision.

References

Aksorn, T. & Hadikusumo, B.H.W. (2008). Critical success factors influencing safety program performance in Thai construction projects. *Safety Science, 46*(4), 709–727.

Al-Humaidi, H. M., & Tan, F. H. (2010). Construction safety in Kuwait. *Journal of Performance of Constructed Facilities, 24*(1), 70–77.

Cheng, E., Fang, D., Li, H., & Xie, F. (2004). Construction safety management: An exploratory study from China. *Construction Innovation, 4*, 229–241.

Choudhry, R. M., Fang, D., & Mohamed, S. (2007). The nature of safety culture: A survey of the state-of-the-art. *Safety Science, 45*(10), 993–1012.

Chua, D. K. H., & Goh, Y. M. (2004). Incident causation model for improving feedback of safety knowledge. *Journal of Construction Engineering and Management, 130*, 542–551.

Dodo, M. (2014). The application of health and safety plan in Nigerian construction firms. *Jordan Journal of Civil Engineering, 8*(1), 81–87.

Hamalainen, P., Takala, J., & Saarela, K. L. (2006). Global estimates of occupational accidents. *Safety Science, 44*, 137–156.

Idubor, E. E., & Oisamoje, M. D. (2013). Management issues in Nigeria's effort to industrialize. *European Scientific Journal, 3*(12), 92–104.

Kalejaiye, P. O. (2013). Occupational health and safety: Issues, challenges and compensation in Nigeria. *Peak Journàl of Public Health and Management, 1*(2), 16–23.

Kartam, N. A., Flood, I., & Koushki, P. (2000). Construction safety in Kuwait: Issues, procedures, problems, and recommendations. *Safety Science, 36*, 163–184.

Kheni, N. A., Dainty, A. R. J., & Gibb, A. (2008). Health and safety management in developing countries: A study of construction SMEs in Ghana. *Construction Management and Economics, 26*(11), 1159–1169.

Kheni, N. A., Gibb, A. G. F., & Dainty, A. R. J. (2006). *The management of construction site health and safety by small and medium-sized construction businesses in developing countries: A Ghana case study*. Paper presented at the 22nd Annual ARCOM Conference, Birmingham.

Kheni, N. A., Gibb, A. G. F., & Dainty, A. R. J. (2010). Health and safety management within small- and medium-sized enterprises (SMEs) in developing countries: Study of contextual influences. *Journal of Construction Engineering and Management, 136*(10), 1104–1115.

Lakovleva, T., Kolvereid, L., & Stephan, U. (2011). Entrepreneurial intentions in developing and developed countries. *Education + Training, 53*(5), 353–370.

Li, R. Y. M. (2013). *Real estate development in developing countries*. Paper presented at the Hong Kong Shue Yan University, Hong Kong.

Li, R. Y. M., & Poon, S. W. (2013). *Construction safety*. Berlin: Springer.

Liao, P. C., Lei, G., Fang, D., & Liu, W. (2014). The relationship between communication and construction safety climate in China. *KSCE Journal of Civil Engineering, 18*(4), 887–897.

National Bureau of Statistics of China. (2013). *Chinese statistical yearbook 2012*. Beijing: China Statistics Press.
Olutuase, S. O. (2014). A study of safety management in the Nigerian construction industry. *IOSR Journal of Business and Management, 16*(3), 1–10.
Snashall, D. (1990). Safety and health in the construction industry. *British Medical Journal, 301*, 563–564.
Tam, C., Zeng, S. & Deng, Z. (2004). Identifying elements of poor construction safety management in China. *Safety Science, 42*(7), 569–586.
The Central People's Government of the People's Republic of China. (2005). *Administrative regulations on the work safety of construction projects*. Retrieved August 22, 2014, from http://www.gov.cn/zwgk/2005-05/23/content_183.htm.
The National Occupational Safety and Health Information Centre. (2006). *Nigeria: Report of the national occupational safety and health information centre*. Retrieved August 1, 2014, from http://www.ilo.org/legacy/english/protection/safework/cis/about/mtg2006/pnga_mlpid.pdf.
The Nigerian Engineering Leadership (NELF). (2013). Retrieved August 25, 2014, from http://www.acen.org.ng/en/images/banners/pdf/reportonlocalcontent.pdf
TheGlobalEconomy.com. (2013). *Country rankings*. Retrieved May 31, 2013, from http://www.theglobaleconomy.com/indicator_static_graph.php?indicatorId=NY.GDP.PCAP.PP.KD&top_ten=0&regions=1w&year=2010.
Umeokafor, N., Isaac, D., Jone, K., & Umeadi, B. (2014). Enforcement of occupational safety and health regulations in Nigeria: An exploration. *European Scientific Journal, 3*, 92–104.
Zeng, S.X., Tam, V.W.Y. & Tam, C.M. (2008). Towards occupational health and safety systems in the construction industry of China. *Safety science, 46*(8), 1155–1168.
Zuo, J., Zillante, G., & Zhao, Z. Y. (2011). Chinese construction industry—Governance, procurement and culture. In S. D. Brunn (Ed.), *Engineering earth*. New York: Kluwer.

Chapter 7
Construction Safety Motivations in Hong Kong: A Psychological Perspective

Rita Yi Man Li and Sun Wah Poon

Abstract There is an increase in the number of construction accidents in recent years in Hong Kong. While most of the previous studies focus on the causes of accidents and the associated costs the research on motivation of safety behaviors among construction practitioners is relative scarce. This chapter sheds light on the findings of a questionnaire survey and the results were analyzed with Principal Component Factor analysis. The findings indicate that recognition of safety behavior and workers' safety concept are good motivation for workers to work safely. Human relationship and goal acceptance are less important factor for safety motivation.

Keywords Construction safety · Motivation · Principal component factor analysis

7.1 Introduction

For decades the safety record in Hong Kong construction industry has remained poor. It records the highest number of fatalities and accident rates among various industry sectors in Hong Kong. Moreover, there is an alarm showing a growing trend in recent years. There were 3,232 industrial accidents in the construction industry, more than 3,160 in 2012 by 2.3 % though lower than 3,833 in 2004 by 15.7 % and higher than the average of the past 5 years 3,029 by 6.7 % (Occupational Safety and Health Branch Labour Department 2014) (Fig. 7.1). Poor safety

R.Y.M. Li (✉)
Department of Economics and Finance / Sustainable Real Estate Research Center, Hong Kong Shue Yan University, Hong Kong, China
e-mail: ritarec1@yahoo.com.hk

R.Y.M. Li
School of Natural and Built Environments, University of South Australia, Adelaide, Australia

S.W. Poon
Department of Real Estate and Construction, The University of Hong Kong, Hong Kong, China

R.Y.M. Li, *Construction Safety and Waste Management*, Risk Engineering, DOI 10.1007/978-3-319-12430-8_7

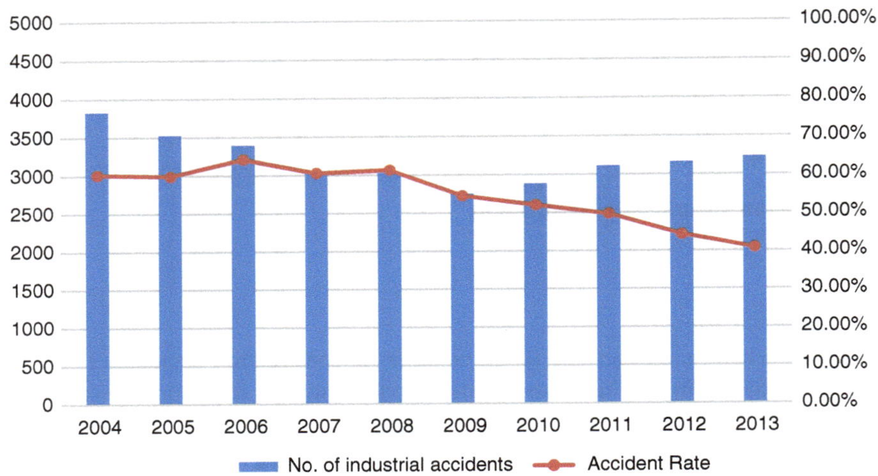

Fig. 7.1 Number of industrial accidents and accident rate per 1,000 workers in construction industry 2004–2013 (Occupational Safety and Health Branch Labour Department 2014)

records impose huge costs in Hong Kong. The employers need to bear the huge monetary and non-monetary costs when accidents occur, for example, medical expenses, compensation payments for pain, lose, suffering and lose, extension of time etc. (Li and Poon 2007, 2013). The contractors may be banned from public works project tender due to poor construction safety record. The injured employees' families suffer from a loss of earning capacity or even loss of their family members (Li and Poon 2007). In view of the worsening safety record in Hong Kong and the huge monetary and non-monetary costs, this chapter reveals factors which affect workers' motivation to work safely from a psychology perspective.

Previous literature shows that humans are motivated by moral acts and approval from others (Sutinen and Kuperan 1999), and behavior may be affected by external factors (McCoy 2014) such as regulation (Li and Ah Pak 2010), pecuniary compensation and promotion (Li and Poon 2011). On the other hand, we may also be motivated intrinsically in absence of reward, i.e. driven by a self-defined goal, social norms and the activity itself (Li and Poon 2011), personal interest (McCoy 2014) self-regulation (Joosten et al. 2009), self-determination, perceptions of challenge and excitement (Li and Poon 2011). Maslow, for example suggests that we are driven by inherent fulfilment rather than peripheral rewards, i.e. extrinsic motivation (Li and Poon 2007).

7.2 Motivation: A Psychology Perspective

"I am content to defend the profit motive but there are also many organizations and individuals who operate within the market economy who are not just motivated by the maximization of gain." (Booth 2011). Motivation is a complex issue as people are driven by a number of different factors. Some of the acts are driven by money but some are not. In general, motivation is a branch of human psychology which studies human acts. The word “motivation” is derived from the Latin word “movere” that means “move” (Qayyum et al. 2011). It is a psychological process and goal-directed action through which people are pushed, persuaded, propelled, energized, directed to engage and sustain in particular behavior (Chovwen and Ivensor 2009; Ghazi et al. 2010; Qayyum et al. 2011). To ensure the success of a company, employers must understand what motivates their employees best as it is essential to improve productivity (Qayyum et al. 2011). As motivation has also been defined as the arousal and persistence of behavior, safety motivation can be defined as arousal and persistence of act which lowers the chance of injury (Li and Poon 2007). In this chapter, we try to perceive the issue from hierarchy of needs theory, two factor theory, expectancy theory, theory, theory Y, Hawthorne investigations and acquired needs theory, and goal setting theory.

7.2.1 Hierarchy of Needs Theory

Maslow’s hierarchy of needs theory suggests that our needs increases from achievement of fundamental physiological needs to the want for self-actualization. Once our basic fundamental needs are satisfied, we pursue higher needs (Zhao and Siu 2014, forthcoming), i.e. satisfying one need sets the stage for moving up to the next (Armstrong et al. 2003). He identifies our wants consist of physiological, safety, love, esteem and self-actualization needs (Li and Poon 2007). Nevertheless, it should be noted that there are different interpretation or version on the Maslow theory of motivation. For example, Liu et al. (2010) suggest that the five basic needs include the lower level physiological, safety and affiliation needs, respect and self-actualization needs are high-level ones. Although Maslow’s theory implies that people need to fulfil the basic survival needs before other issues such as safety, empirical observations have found that such step-by-step, well-organized scale up needs may not happen in reality (Li and Poon 2007).

7.2.2 Two Factor Theory

As Frederick Herzberg finds that Maslow’s motivation theory is imperfect, he modifies it to “two factor theory” (Li and Poon 2007). The two factor theory highlights the factors which affect our behavior as inhibitors (also known as hygiene and demotivators) and enablers (also called as motivators and satisfiers) (Parka and Ryoo 2013). The idea is built according to the responses of 4,000 people. Many of them identify job nature as the factor that motivates them to do the work. Herzberg calls it “satisfier factor”. When they are asked what deters them to do the work, many of them identify them as work setting. Herzberg calls them "hygiene factors". While high level of "satisfier" leads to greater level of motivation, hygiene factors only affect motivation up to a certain level if and only if there is some deficiency in the present environment (Li and Poon 2007). Hygiene factors do not motivate us when they are there but in absence of them will lead to decrease in motivation. For example, the presence of lighting will not motivate us to work but absence of that will lead to a lower work motivation. Besides, it is found that the presence of motivation factors not only increase our motivation but also level of satisfaction (Li and Poon 2013). Nevertheless, Herzberg's idea is also criticized about the applicability of non-professional groups e.g. construction workers (Li and Poon 2007).

7.2.3 Expectancy Theory

Expectancy theory has been regarded as one of the most widespread ways which determine individual motivation (Liao et al. 2014). Expectancy theory has a close relationship with incentive systems. It rewards those who have worked hard. While Maslow suggests that our achievement is largely related to the need for inherent fulfillment rather than the need for peripheral rewards which include money. Vroom proposes motivation is determined by the perceived relationship between our action and the results. The force which motivates us is a multiple of instrumentality, valence and expectancy:

1. Expectancy concedes that we are of the view that performance perfection requires sweats and tears of our hard work.
2. Instrumentality is the scope to which improved performance will lead to a specific outcome.
3. Individual Valence implies the more stalwartly an individual recognizes that his performance will lead to an outcome, the more motivated that individual will be (Li and Poon 2007).

In the same vein, Liao et al. (2014) suggests that the amount of effort exerted by us according to expectation theory is based on a systematic analysis of:

1. The values of the rewards arise from outcomes.
2. The chance that the rewards will result from the outcomes.
3. The probability of attaining these outcomes via their efforts and actions (Liao et al. 2014).

Expectation theory implies that successful safety policy should minimize the possible side effects. For example, safety measures which lead to an increase in time costs to workers, tempt them to take risk instead. The theory echoes the economic theory that people weigh costs and benefits before they make decisions. Although expectancy theory presents a broad-spectrum structure for motivation behavior at work, there is a lack of construct validity of mechanism of the theory (Li and Poon 2007).

Table 7.1 Viewpoints on punishment and incentives which motivate safety behavior

Author	Viewpoints on incentives and punishment which lead to safe behaviour
Peters (1991)	A reduction in construction accidents and injuries when there are safety incentives
McAfee and Winn (1989)	Safety indices did not get better in spite of the provision of safety incentives
Shields (1995)	Incentive programs produce too few winners. This can impel distrust and create corruption which crop up even greater concern. Moreover, employees may hide near misses and minor incidents which aggravates future program. This approach has limited success as motivation is short-lived
Goetsch (2003)	Incentives can improve safety performance given that the incentives are used properly

7.2.4 Theory X and Y

Theory X assumes that we dislike work and try to get rid of it. Hence, we must be directed and threatened with punishment. Theory Y represents another side of the coin about human behaviors. It assumes that work is as natural as rest and we all prefer to take responsibility. External control and punishment are not the only means which lead to an increase effort on organizational objective. We are motivated when we are cherished as a precious member and suitable working environment is provided (Li 2009; Li and Poon 2007). When theory X is put to practice, safety officers penalize workers who fail to comply with safety requirements. If they are followers of theory Y, incentive will be given to workers. Nevertheless, these two theories represent two extremes that were unusual in practice (Li and Poon 2007). Hertzberg, for example, suggests that we are motivated by a combinations of positive rewards and negative punishment (Bierma and Krishnan 1997).

Other writers' viewpoints on safety and punishment are listed in Table 7.1.

7.2.4.1 Safety Incentives and Punishment

When an unsafe act results in immediate positive reinforcement and potential punishments are not intense, delayed and irregular, people choose to get hurt. Punishment is unlikely to be effective if it is not frequent and of mild intensity. The use of penalization has shown limited success, largely for the reason that punishment is time and again apprehended to be less effective than positive reinforcement. Besides, foremen are fear of resentment which may lead to lack of collaboration and output (Li and Poon 2007).

7.2.5 Hawthorne Investigations

The Hawthorne Studies are one of the largest and the most influential investigations in organizational research done by Elton Mayo and his research team. The research results highlight the importance of human in a company and reinforce the idea that as the Company grows, the role of human being will become more important. The theory, however, has been criticized as the research results are "closed-system" analysis. It tells little about climate or culture within which the organization operate (Hassard 2012). Elton's research team finds that psychological instead of physical factors motivate us. The theory suggests that good working environment is not enough to develop a safety work environment in construction industry. Informal group culture may affect workers' safety behavior (Li and Poon 2007).

7.2.6 Acquired Needs Theory

David McClelland's research suggests that we desire for power to control others' behavior. In view of this, if the employers want to enhance construction practitioners' safety motivation, power should be given to their workers. Most of the social behavior can be understood in terms of an interaction between underlying motives and features of the environment (Li and Poon 2007). He further suggested that there are three basic motives as follows:

1. Achievement motivation (nAch): the driving force to do concrete tasks (Yagil 2006) from our mind, i.e. when we are committed to pursuing achievement goals and a standard of excellence on own initiatives. This does not, however, mean that the goals set by ourselves should lead to an achievement motivation which is more important than goals given by the others others. nAch is also defined as the desire to do things as rapidly/good as possible and overcome obstacles, attain high standards, excel ourselves and surpass others. Generally speaking, the achievement motive can be classified into avoidance tendency "fear of failure" and approach tendency "hope of success" (Sandalgaard et al. 2011).
2. Affiliation motivation (nAffi): the desire to be affiliated with other people (Yagil 2006). It is aroused by the need of being affiliated and liked to others. It concerns with restoring, maintaining, and establishing a positive affective relationship with the others (Sandalgaard et al. 2011).
3. Power motivation (nPow): the desire to influence others (Yagil 2006). This motive can be further divided into hope and fear of power (Sandalgaard et al. 2011). Those who are motivated by strong power may express this motive via various socially acceptable forms of controlling and influencing others via service provision and helping the others (Yagil 2006).

7.2.7 Goal Setting Theory

Developed by Locke in 1968, Goal setting theory of motivation holds that intention or goals are the most powerful determinant of our behavior. Goal setting affects performance by boosting motivation, mobilizing effort, directing the attention and actions of us. Award and incentive offered us should elicit better performance level as it causes the individual to change our goal and intention. Goal setting theory accepts that goals are the immediate regulators of our action. Our performance will be improved when the goals are accepted by us and if the goals are effective and higher goals lead to better performance than lower goals (Li 2006). Therefore, effective goal setting can affect organizational performance (Jung 2014).

7.3 Application of Psychological Theories on Construction Safety Motivation

In the research conducted by Li (2006), 42 effective questionnaires were received. Participants were asked 14 questions (Table 7.2) on motivation based on 5-point-likert scale (from 1 Strongly Disagree to 5 Strongly Agree) and the results are listed in the following Table. As there are a number of unobserved latent variables which account for the correlations among the observed variables, such that each observed variable (y) can be expressed as a weighted composite of a set of latent variables (F_s).

$$Y_i = I_1F_1 + A_{I2}F_2 + \cdots + A_{ik}F_k + e_i$$

Y_i is the ith observed variable on the factors and e_i is the residual of Y_i on the factors. Given the assumption that residuals are not correlated among the observed variables, correlations among the observed variables are accounted for by the factors. Factor analysis in the first stage produces an initial orthogonal solution and the second stage produces a rotated solution from the solution obtained at the first stage (Li and Poon 2007) (Table 7.3).

Eight factors of motivation are extracted after VARIMAX rotation as factor 1 which include bonus, penalty, awards, power, achievement, recognition, advancement, work conditions. Since all these factors are related to recognition of the workers safety behavior, factor 1 is named as "recognition of the workers safety behavior". Factor 2 consisted of Safety behavior achieving desired level of safety performance, safety work outcomes and the priority of safety need. These 3 factors are about the safety concept of workers. Therefore factor 2 is called "the safety concept of workers". Factor 3 consists of 2 factors: workers in the group and relationships with peers. These 2 factors concerned about human relation and was then renamed as "the human relationship". Factor 4 is goal acceptance of safety by individuals.

Table 7.2 Fourteen questions about safety motivations in the questionnaire (Li 2006)

Items of motivation	Abbreviation	Name of the psychological theory	Average score our of 5	Ranking
1. Bonus is my safety motivation	Bonus	Theory Y	4.548	1
2. Safety behavior achieves desired level of safety performance	Beha	Expectancy theory	4.143	2
3. Awards are my safety motivation	Award	Theory Y	3.857	3
4. Safety goals' acceptance by individual is my safety motivation	Goalacc	Goal setting theory	2.857	4
5. Penalties are my safety motivation	Penal	Theory X	3.857	5
6. Power is my safety motivation	Power	Acquired needs theory	3.810	6
7. Achievement is my safety motivation	Achieve	Herberg/ acquired needs theory	3.786	7
8. Recognition is my safety motivation	Recognition	Two factor theory	3.762	8
9. Advancement is my safety motivation	Advancement	Two factor theory	3.738	9
10. The safety work outcomes are my safety motivation	Outcome	Expectancy theory	3.667	10
11. Relationship with peers is my safety motivation	Peers	Two factor theory	3.643	11
12. The priority of safety need compare to other needs such as salary	Prior	Hierarchy of needs theory	3.595	12
13. Work conditions are my safety motivation	Wkcon	Two factor theory	3.5	13
14. Workers in the group can affect my safety behavior	Wkgp	Hawthorne investigations	3.105	14

The effect of these factors was ranked by the formula $F_i = \sum_{j=1}^{n} \frac{A_{ij}}{n}$ where F_i is the factor score, A_{ij} is the mean score of the jth factor of safety motivation.

Factor 1 = (scoring of workers in work group + scoring of relationship with peers)/(number of factor, i.e. 2 in factor 1).
Factor 1 = 3.1048 + 3.6429/2 = 3.3739 (Li and Poon 2007) (Table 7.4).

Results of our research indicate that recognition of safety behaviors and safety concept of construction practitioners are good motivation for workers to work safely. Human relationship and goal acceptance are less important safety factor of motivation.

Table 7.3 Results of VARIMAX rotation (Li 2006)

Component	Factor 1	Factor 2	Factor 3	Factor 4
Achieve	0.873	0.035	−0.372	0.001
Advance	0.816	0.328	0.192	−0.064
Award	0.92	0.207	0.126	−0.047
Beha	0.283	0.586	0.528	−0.026
Bonus	0.875	0.117	−0.406	−0.122
Goalacc	−0.114	0.039	−0.088	0.936
Outcome	0.183	0.886	0.093	−0.155
Penal	0.781	−0.22	0.303	−0.351
Power	0.822	0.196	0.258	−0.089
Prior	−0.025	0.826	−0.071	0.156
Reco	0.84	0.165	0.115	0.225
Relat	0.22	−0.179	0.684	0.548
Wkcon	0.865	−0.188	0.12	0.023
Wkgp	−0.02	0.137	0.873	−0.184
Percentage of variance	42.53	15.55	14.81	10.44

Table 7.4 Factor score of 4 factors (Li 2006)

Name of the factor	Components of the factor	Factor score	Rank
Safety behaviour recognition	Bonus, penalty, awards, power, achievement, recognition, advancement, work conditions	3.86	1
Workers' safety concept	Safety behavior achieving desired level of safety performance, safety work outcomes and the priority of safety need	3.80	2
Human relationship	Workers in the group and relationships with peers	3.37	3
Goal acceptance of safety behavior	Goal acceptance of safety by us	2.86	4

7.4 Conclusions

This chapter reviews the factors which motivate workers to work safely according to psychological theories. The results of the questionnaires show that the safety behavior recognition through bonus, penalty, awards, advancement etc. is the most important factor which motivates us to work safely. The goal acceptance of safety behavior is of the least important factors among the four major factors. It suggests that extrinsic motivation is of stronger force than intrinsic motivation.

References

Armstrong, S., Appelbaum, M., & Henches, K. (2003). *Stress-free performance appraisals: Turn your most painful management duty into a powerful motivational tool*. New York: The Career Press.

Bierma, T., & Krishnan, U. (1997). Educator-employer partnerships: A tool for improving environmental health education. *Journal of Environmental Health, 60*(1), 11–15.

Booth, P. (2011). The Bishop of Brentwood, migrants, the market and the pursuit of gain. *Economic Affairs, 31*(3), 49–50.

Chovwen, C., & Ivensor, E. (2009). Job insecurity and motivation among women in Nigerian consolidated banks. *Gender in Management: An International Journal, 24*(5), 316–326.

Ghazi, S. R., Ali, R., Shahzad, S., Khan, M. S., & Hukamdad, H. (2010). Parental involvement in children academic motivation. *Asian Social Science, 6*(4), 93–99.

Goetsch, D. L. (2003). *Construction safety and health*. Upper Saddle River, New York: Prentice Hall.

Hassard, J. S. (2012). Rethinking the Hawthorne studies: The western electric research in its social, political and historical context. *Human Relations, 65*(11), 1431–1461.

Joosten, A. V., Bundy, A. C., & Einfeld, S. L. (2009). Intrinsic and extrinsic motivation for stereotypic and repetitive behavior. *Journal of Autism and Developmental Disorders, 39*, 521–531.

Jung, C. S. (2014). Why are goals important in the public sector? Exploring the benefits of goal clarity for reducing turnover intention. *Journal of Public Administration and Research Theory, 24*(1), 209–234.

Li, R. Y. M. (2006). *Effectiveness of various construction safety measures in Hong Kong*. Unpublished B.Sc. thesis, The University of Hong Kong, Hong Kong.

Li, R. Y. M. (2009). Achieving compliance with environmental health-related land use planning conditions in Hong Kong: Perspectives from traditional motivation theories. *Journal of Environmental Health, 72*(4), 22–25.

Li, R. Y. M., & Ah Pak, D. H. (2010). Resistance and motivation to share sustainable development knowledge by Web 2.0. *Journal of Information and Knowledge Management*, *9*(3), 251–262.

Li, R. Y. M., & Poon, S. W. (2007). *A critical review of construction safety motivation in Hong Kong*. Paper presented at the Management Science and Engineering Academic Conference, Tianjin.

Li, R. Y. M., & Poon, S. W. (2011). Using Web 2.0 to share the knowledge of construction safety as a public good in nature among researchers: The fable of economic animals. *Economic Affairs, 31*(1), 73–79.

Li, R. Y. M., & Poon, S. W. (2013). *Construction safety*. Germany: Springer.

Liao, H. L., Liu, S. H., & Pi, S. M. (2014). Modeling motivations for blogging: An expectancy theory analysis. *Social Behavior and Personality: An International Journal, 39*(2), 251–264.

Liu, X., Bao, Z., & Wang, Z. (2010). Internet use and internet addiction disorder among medical students: A case from China. *Asia Social Science, 6*(1), 28–34.

McAfee, R. B., & Winn, A. R. (1989). The use of incentives and feed-back to enhance work place safety: A critique of the literature. *Journal of Safety Research, 20*, 7–19.

McCoy, D. C. (2014). Student motivation for learning in Ghana: Relationships with Caregivers' values toward education, attendance, and academic achievement. *School Psychology International, 35*(3), 294–308.

Occupational Safety and Health Branch Labour Department. (2014). Occupational Safety and Health Statistics Bulletin.

Parka, S. C., & Ryoo, S. Y. (2013). An empirical investigation of end-users' switching toward cloud computing: A two factor theory perspective. *Computers in Human Behavior, 29*, 160–170.

Peters, R. H. (1991). Strategies for encouraging self-protective employee behaviour. *Journal of Safety Research, 22*(2), 53–70.

Qayyum, A., Mahmood, S., & Mahmood, A. (2011). A preliminary investigation of employee motivation in Pakistan's banking sector. *Research and Practice in Human Resource Management, 19*(1), 38–52.

Sandalgaard, N., Bukh, P. N., & Poulsen, C. S. (2011). The interaction between motivational disposition and participative budgeting. *Journal of Human Resource Costing and Accounting, 15*(1), 7–23.

Shields, M. A. (1995). Human relations in safety. *Safety Bulletin, 12*(1), 6–7.

Sutinen, J. G., & Kuperan, K. (1999). A socio-economic theory of regulatory compliance. *International Journal of Social Economics, 26*(1/2/3), 174–193.

Yagil, D. (2006). The relationship of service provider power motivation, empowerment and burnout to customer satisfaction. *International Journal of Service Industry Management, 17*(3), 258–270.

Zhao, T., & Siu, K. W. M. (2014, forthcoming). The needs for quality urban rail transit life in Asian Metropolitan Cities. *Applied Research in Quality of Life.*

Chapter 8
A Conceptual Study of Construction Workers' Safety Performance from Safety Climate and Social Exchange Perspectives

Rita Peihua Zhang and Rita Yi Man Li

Abstract Construction safety performance can be measured in a number of ways. Some of the popular indicators include number of construction accidents, fatalities and fatalities rate and compensation. Nevertheless, it is criticised that these lagging indicators cannot accurately measure safety performance on sites. Hence, another strand of literature views the issue from safety climate and social exchange perspective. Safety climate, our perceptions of safety practices, policies, procedures about safety on site provides an indicator for safety priority at work. On the other hand, social exchange theory perceived organisation support affects our obligation, gratitude and trust on various issues in workplace including safety. Perceived organizational support and leader membership exchange, under the umbrella of social exchange, are two major items which may affect construction safety performance.

Keywords Safety performance · Safety climate · Social exchange

8.1 Introduction

The role of safety climate in safety performance at work has been well-informed by numerous studies in different industries and countries. For example, Williamson et al. (1997), Flin et al. (2000), Fang et al. (2006) validate the dimensions underlying safety climate construct. Mearns et al. (2003), Cooper and Phillips (2004),

R.P. Zhang
School of Property, Construction and Project Management, RMIT University, Melbourne, Australia

R.Y.M. Li (✉)
Department of Economics and Finance / Sustainable Real Estate Research Center, Hong Kong Shue Yan University, North Point, Hong Kong
e-mail: ritarec1@yahoo.com.hk

R.Y.M. Li
School of Natural and Built Environments, University of South Australia, Adelaide, Australia

R.Y.M. Li, *Construction Safety and Waste Management*,
Risk Engineering, DOI 10.1007/978-3-319-12430-8_8

Table 8.1 Types of safety performance measurement methods (authors' Table)

Author	Types of safety performance	City, country
Irumba (2014)	Construction accidents	Kampala, Uganda
Suárez-Cebador et al. (2014)	Construction accidents	Spain
Li and Poon (2013)	Construction accident compensation; construction accidents	United Kingdom
Li and Poon (2009a, b)	Construction accident compensation	Hong Kong
Hinze and Teizer (2011)	Fatalities	United States
Fabrega and Starkey (2001)	Fatalities	United States
Dement and Lipscomb (1999)	Construction accident compensation	United States
Ling et al. (2009)	Fatalities	Singapore

Smith et al. (2006) and Clarke (2006a, b) empirically test the relationship between safety climate, safety-related behaviours and outcomes. Zohar (2008), Meliá et al. (2008), Lingard et al. (2010a, b), Brondino et al. (2012), Lingard et al. (2012) develop multi-level safety climate measures to distinguish between organizational level safety climate and group level safety climate. Based on 30 years' of safety climate research, Zohar concludes that it is time to move to 'the next phase of scientific enquiry in which constructs are being augmented by testing safety climate relationships with antecedents, moderators, and mediators, as well as relationships with other established constructs' (Zohar 2010, p. 1517) (Table 8.1).

Researchers suggest that social exchange relationship between employees, leaders, and organizations may have important safety implications (Hofmann and Morgeson 1999). Social exchange engenders individuals' personal obligation, gratitude, trust and is characterized by reciprocity (Blau 1964). In light of social exchange principle, one party's favour would lead to another party's positive action in return. For example, when an organization shows an extended consideration for employees' well-being and investments on employees' career, employees are found to reciprocate with improved performance on core tasks, high level of organizational citizenship behaviour, more affective commitment and intention to stay (Tsui et al. 1997). Social exchange relationship is also applicable in safety management, i.e. if employees perceive that management emphasize on the importance of safety, and care of employees' safety and well-being, employees would feel obliged to reciprocate with compliance of safety rules and procedures, as well as safety citizenship behaviour. Recently, a number of empirical studies have been conducted to explore the linkage between safety climate and safety-related behaviours/outcomes by integrating social exchange perspective. For example, Kath et al. (2010) report that railway workers' upward safety communication behaviour (i.e. discussing safety issues with their supervisors) is jointly determined by safety climate predictors and social exchange relationship between workers and their supervisors, which is known as leader-member exchange (LMX). DeJoy et al. (2010) empirically test a model of social exchange and safety management, which mainly focuses on the exchange relationship between organization and employees, i.e. known as perceived organizational support (POS). They find that POS affect employees' perceptions of organizational safety climate, which in turn affect perceived safety at work and self-reported accidents. The effect of social exchange in shaping employees' safety-related attitudes and behaviours implies that safety outcomes are part of the dynamic interactions between employers and employees (DeJoy et al. 2010).

Safety performance in the construction industry is relatively poor as compared to other industries. A large number of construction injuries and fatalities have been reported worldwide (Atkinson and Westall 2010). There has been an increasing consensus that organizational, managerial, and human factors are critical causes of accidents rather than pure technological failures in construction industry (Mohamed 2002). This has driven researchers to consider originating managerial and culture factors that shape acts and conditions in work places (Behm and Schneller 2012; Haslam et al. 2005). Fang et al. (2006), Meliá et al. (2008), Lingard et al. (2009), Lingard et al. (2010a, b), Lingard et al. (2012). Substantial work has also been carried out by researchers to measure safety climate in the construction industry to diagnose the difficulties or problems in safety management systems). The construction industry is operated around projects, and workers from the same construction company are assigned to different workgroups for different projects. The construction sites are usually spatially remote from the company office, resulting in few opportunities for on-site workers to directly interact with the senior management of their company. However, they have daily communications with their group supervisors, whose supervisory safety leaderships give rise to the formation of group-level safety climates. Previous studies have provided empirical evidence that multi-level safety climates exist in the construction industry (Lingard et al. 2009, 2010a, b, 2012; Meliá et al. 2008). However, most of the existing work only focuses on the effect of the multi-level safety climates on workers' safety behaviour or direct safety outcomes. It is anticipated that the exchange relationships between workers, construction organizations, and supervisors also have important implications on worker's safety performance. This paper aims to develop a conceptual explanation of the predictors which affect construction workers' safety performance by integrating safety climate and social exchange perspectives, and propose an integrative research model for future empirical testing.

8.2 Safety Performance

A comprehensive literature review indicates that existing literature mainly focuses on the direct safety outcomes: recorded accident injuries, compensation (Dement and Lipscomb 1999) accident rate, and number of fatalities/fatalities rate (Fabrega and Starkey 2001; Hinze and Teizer 2011; Ling et al. 2009). Dement and Lipscomb (1999)'s research shows that there were 31,113 workers' compensation claims among 7,400 North Carolina Homebuilders Association members and their subcontractors from 1986 to 1994. Tam et al. (2006) identify accident per 1,000 workers as the safety performance. Ling et al. (2009)'s research shows that the fatality rate in Singapore's construction industry is unacceptably high as 39 % of the total workplace fatalities come from construction sector alone. Waehrer et al. (2007) uses national incidence data from the Bureau of Labor Statistics to construct a comprehensive cost model which includes indirect losses in wage, direct medical costs and household productivity.

Neal et al. (2000), however, note that there has been a shift in safety literature from individual level factors which lead to incidents and accidents such as non-compliance with safety procedures and human error or to organizational factors which include safety climate in recent years. It is argued that these 'lagging indicators' are not able reflect workers' safety attitude and safety behaviour. Griffin and Neal (2000) and Neal et al. (2000) point out that one of the key assumption of studies which examines the relationship between safety climate and direct safety outcome is that individual safety performance plays the mediating role. In existing literature, however, not many studies have explicitly examined the influence of safety climate on workers' safety performance. It is worthwhile to explore how safety climate predicts safety outcomes through its shaping effect on individual's safety attitude and behaviour.

Marchand et al. (1998) argue that the traditional way of measuring safety behaviour is too limited as it only focuses on employees' compliance to safety rules. They suggest that workers' initiatives should also be taken into account when developing safety behaviour measures. In Griffin and Neal (2000), Neal et al. (2000), Neal and Griffin (2006)'s studies, safety compliance and safety participant are found to be the two major components which affect safety performance. These two components were developed based on Borman and Motowidlo's (1993) idea that performance consists of task performance and contextual performance. Task performance is 'the activities that are formally recognized as part of their jobs, activities that contribute to the organization's technical core either directly or indirectly'; while contextual performance 'supports the organizational and psychological environment in which the technical core must function' (Borman and Motowidlo 1993, p. 73). According to these definitions, safety compliance is used to describe the mandatory core safety activities that individuals need to perform to maintain workplace safety (Griffin and Neal 2000). Examples of workers' safety compliance may include following correct working procedures, adhering to safety rules and wearing appropriate personal protection equipment (PPE). Safety participant is considered to be more voluntary in nature, and includes behaviours that are beyond workers' formal duties, e.g. safety extra-roles or safety citizenship behaviours. Examples of safety participant may include looking after co-workers' safety, promoting safety program in workplace, demonstrating initiatives in improving safety in workplace (Neal et al. 2000). These behaviours help to create a psychological environment that supports safety.

The two component safety performance model has recently received much recognition and application in safety management research. For example, Vinodkumar and Bhasi (2010) found that some safety management practices influence safety compliance and safety participation through the mediating effect of workers' safety knowledge and motivation. In a comprehensive meta-analysis conducted by Clarke (2006a, b), safety compliance and safety participation are found to partially mediate the effect of safety climate on accident involvement. This study will also examine the two component safety performance as dependent construct and identifies predictors to construction workers' safety performance.

8.3 Organisation Climate, Safety Climate and Safety Performance

Neal et al. (2000) suggest that organizational climate is a multidimensional construct which encompasses a wide range of individual evaluations of our work environment. It exerts strong impact on our motivation to achieve some desire outcomes. Evaluations on organisational climate may focus on general dimensions of the environment which includes communication, leadership. General perceptions of the organizational context affect the interactions among us, attitudes on organizational rewards and responses to changes in work environment. General organizational climate also affect the skills and knowledge by increasing participation in various work related activities such as training. On the other hand, specific dimensions include customer service climate or safety climate.

Under the lens of Neal et al. (2000), safety climate describes our perceptions on the value of safety in work environment. A range of factors have been identified as important components in safety climate:

1. Management values, for example, management concern for employees' well-being;
2. Management and organizational practices, for example, adequacy of training, provision of safety equipment, quality of safety management systems;
3. Communication and workers' involvement in workplace health and safety (Neal et al. 2000).

Previous research shows that strong and positive safety climate are linked to high levels of safety performance (Lingard et al. 2011). In general, safety climate is defined as 'individual perceptions of the policies, procedures, and practices relating to safety in the workplace' (Neal and Griffin 2006). It provides an indication of 'priority of safety' in an organization relative to other priorities such as production and cost (Zohar 2000). Safety climate shapes worker's safety behaviour through formed expectations about how safety is valued and rewarded by organizations (Lingard et al. 2010a, b). Therefore, the behaviors and safety climate of site managers and workers are crucial (Holt 2001). A strong and positive connection between safety climate and safety behaviours/outcomes has been supported in a large number of studies. For example, Clarke (2006a, b) find that a positive safety climate, which is characterized by workers' positive response to safety and low conflict between production and safety, significantly predicts workers unsafety behaviour in a

UK-based car manufacturing plant. Based on a survey conducted in four industries of manufacturing, construction, service and transportation, Huang et al. (2006) reported that positive safety climate is reversely associated with employees' self-reported occupational injury and the relationship is mediated by employees' safety control. Neal and Griffin (2006) reported a lagged effect of safety climate on individuals' safety behaviour. Specifically, safety climate at one time predicts subsequent changes in employees' safety motivation, which in turn leads to subsequent changes in employees' safety behaviour. Employees' improved safety behaviour then positively associate with a subsequent decrease in accidents. In the construction industry, strong evidence is also provided by researchers regarding the significant impact of safety climate on safety performance. For example, in the US construction industry, Gillen et al. (2002) discover that the injury severity reported by injured construction workers is significantly correlated to the workers' safety climate perceptions in terms of management's concern with their well-being, managements' actual respond to this concern as well as existing physical working environment condition. Siu et al. (2004) find that workers' perceptions of the safety attitudes hold by their colleagues, mangers, safety officers, supervisors as well as themselves predict worker's self-reported occupational injuries in the Hong Kong construction industry.

Safety climate in early safety management studies mainly refers to organizational safety climate. Although employees' perceptions of their supervisors' and co-workers' safety attitudes and behaviours are also assessed occasionally, they are considered to be part of the organizational safety climate. There is no specific differentiation between organizational-level and group-level climate in these studies. Researchers argue that a more fine-grained safety climate analysis is required in large and complex organizations such as the construction industry, where decentralized organizations and non-routine works are involved (Lingard et al. 2009, 2011). Although recent academic interest in measurement of safety climate has resulted in a growth in assessment instruments via large-scale surveys, there is no consensus on theoretical framework in this area (Flin et al. 2000). Construction workers from the same construction companies are assigned to different workgroups for different construction projects. These workgroups are usually spatially dispersed from each other and remote from their company offices. Construction workers usually have few opportunities to directly communicate with top management of their organizations. However, they have daily communications with their supervisors, who provide daily instructions and guidance to their work. According to Zohar (2000), top management are mainly concerned with policies making and establishment of procedures for policy implementation, while supervisors are those who execute the procedures by turning them into context-specific action directives. This gives rise to the source of safety climate perceptions for two levels of analysis, i.e. policies and procedures related to organizational level and supervisory practice related to group level (Zohar 2000). Supervisors play a critical role in conveying safety message from senior management level to workforce level (O'Dea and Flin 2001). Due to supervisors' discrepant interpretation and implementation of formal procedures, different group-level climates are created. In the study conducted by Lingard et al. (2009), it is empirically found that distinct workgroup safety climates exist in the construction industry, which explains different safety performance among workgroups.

Employees form their perceptions of group safety climate mainly by interpreting whether supervisory practices converge into an internally consistent pattern regarding relative priority of safety over competing operational goals such as speed and cost (Zohar 2000; Zohar and Luria 2005). For example, it is considered to be an inconsistent pattern if a supervisor emphasize safety in one occasion while ignores safety in another situation. Workers' perception of supervisors' safety priority, to certain extent, reflects supervisors' commitment to safety. Prior research has provided substantial evidence that supervisory safety leadership plays an important role in safety management. For example, Zohar (2000) find that group climate reflected by workers' perceptions of supervisory practice significantly predicts micro-accident record in manufacture sector. Lingard et al. (2010a, b) reveal that high level of supervisors' safety expectation is one important facet of strongly supportive safety climate for workgroups in the construction industry. They discover that workgroups with strong supportive safety climate report lower injury frequency rates than other workgroups. Therefore, it is hypothesized as below.

8.4 Social Exchange, Perceived Organisation Support, Leader Member Exchange and Safety Behaviour

8.4.1 Social Exchange Theory (SET)

Social exchange theory (SET) suggests that our behaviour is a kind of reciprocation relationship (Mansor and Ali 2010), depend on the perceived rewards, i.e. the perceived benefits that will bring to us (Tyrie and Ferguson 2013; Zhang and Ng 2012). We act accordingly to maximize our benefits and minimise the costs. If we do not receive any reward after providing favour to others, we will not do them a favour later. On the other hand, if the other reciprocate with a return, more rounds of exchanges will become possible (Zhang and Ng 2012). This action is mainly driven by the one who receive benefits from others later feel that there is an obligation to compensate through effort and loyalty (Mustapha et al. 2011). The idea is similar to n-rounds game theory where participants' behaviour in stage 1 will affect the later stage (Li 2011; Li and Poon 2011). Hence, SET is characterised by reciprocal relationship where our action will affect the others (Zhang and Ng 2012). Cook et al. (2013) concedes that social exchange is an exchange activity of tangible or intangible, rewarding or costly materials between at least two parties according to the function of payoff. Therefore, Skinner's enforcement principles derived from behaviourism can be used to explain the persistence of exchange relations (Cook et al. 2013), some of the motivations theory such as intrinsic and extrinsic motivation, theory X, theory Y etc. are valuable to identify the usefulness of external forces (rewards, punishment etc.) or internal forces (e.g. morally driven by ourselves) (Li 2009; Li and Poon 2009a, b) in these exchange activities.

8.4.2 Perceived Organizational Support (POS)

Under the framework of social exchange, Eisenberger et al. (1986) argue that employees' commitment to an organization is strongly depended on their perception of the organization's commitment to them. The social exchange interpretation of organization's commitment is conceptualized into Perceived Organizational Support (POS), which represents employees' 'global beliefs concerning the extent to which the organization values their contributions and cares about their welling-being' (Eisenberger et al. 1986, p. 501). Eisenberger and colleagues suggest that employees form the beliefs by evaluating various aspects of treatments they receive from the organization, and such beliefs provide the basis for employees to make inference about the organization's 'readiness to reward increased effort and to meet needs for praise and approval' (Eisenberger et al. 1986, p. 501). More specifically, employees who perceive high level of POS would have high expectancy that their greater work effort would be rewarded and they would have a stronger attachment to the organization. The effort-outcome expectancy and affective attachment will lead employees to show more affective commitment to their organizations and have more involvement in organizational activities to support the organizational goals (Wayne et al. 1997). Prior studies have also shown that POS contribute to employees' positive attitudes and behaviours, including increased commitment, less intention to leave, reduced absenteeism, improved performance, etc. (Eisenberger et al. 1986; Michael et al. 2005; Settoon et al. 1996; Wayne et al. 1997).

Hofmann et al. (2003) suggest that the type of behaviour that is valued in a specific work environment provides employees with the direction for reciprocation. In some industries where safety is highly concerned (e.g. manufacture, construction, mining), employees would reciprocate high level POS with improved safety performance. Due to that POS reflects organization's general traits, the link between POS and specific safety outcomes is usually found to be mediated by constructs related to specific safety attitude and perceptions. For instance, Hofmann and Morgeson (1999) report that POS reduces accidents through the mediating effect of safety communication and safety commitment in the manufacture sector. In another study, Wallace et al. (2006) suggest safety climate to be the mediator between perceived organizational support and accidents. Based on Bowen and Schneider's (1993) work, Wallace and colleagues argue that POS reflects foundation climate, which represents employees' perception of the more encompassing environment in an organization; foundation climate may not strongly correlate to specific outcome, instead it affects specific organizational outcomes through specific climate, which reflects employees' perception of a particular area of interest such as safety. They suggest that the mechanism behind the general-specific-outcome framework is that foundation climate establishes certain norms, and the norms are further solidified in more specific climate which is directly related to outcomes. In a high risk working environment where safety is one of the main concerns, employees may perceive POS partly by evaluating organization's effort to ensure employee's safety. High level POS may suggest employees' perception that their organization cares about their wellbeing, and devote effort to ensure workers' health and safety in workplace. This, in turn, may influence employees' perception of safety climate in the organization. Thus, it is hypothesized that as below.

8.4.3 Leader Member Exchange (LMX)

Another construct frequently examined in organization studies is leader-member exchange (LMX), which indicates the quality of relationship between an employee and his/her supervisor. LMX is based on the leadership theory of Vertical Dyad Linkage (VDL) Model, which assumes that leaders differentiate among their subordinates regarding leader behaviours (Liden and Graen 1980). Although the supervisor is one of the organization's representatives/agents, the employee—supervisor exchange is distinctive from the employee—organization exchange. An organization is an entity with a collective of many individuals. The organization's action is represented by the actions of its agents (e.g. directors, managers) which is usually based on historical decisions. When an employee exchanges with an organization, he/she doesn't develop a relationship with any individuals. With the existence of high level POS, the employee would be obligated to reciprocate with attitudes and behaviours that benefit the general organization, such as stronger sense of identity, more commitment to stay with better job performance. Regarding the leader-member exchange, the employee develops a relationship with a specific individual, i.e. his/her supervisor. If the employee receives cares and support from the supervisor, he/she would be obligated to reciprocate with actions that directly benefit the supervisor. Gouldner (1960) suggests that reciprocity in social exchange creates social stability, goes beyond the compliance of specific social roles, e.g. scope of formal job duties defined in a contract. Supervisors and group workers perform specific job duties regardless of the quality of their relationships. However, with high quality leader-member exchange relationship, employees not only comply to their formal job duties, but also engage in behaviours that are beneficial to supervisors, work beyond the scope of formal job duties (Wayne et al. 1997).

In working environments where safety is concerned (e.g. construction site), employees may reciprocate high quality LMX with compliance with safety requirements as well as safety related behaviours, such as an openly raise safety concerns to their supervisors (Hofmann and Morgeson 1999), monitoring co-workers' safety behaviour, correct potential safety problems and report dangers (Mearns and Reader 2008). These behaviours will ultimately help improve workplace safety performance and reduce accidents. In high risk working environments, safety record is usually used as an indicator to evaluate supervisors' performance. Employee's reciprocation of safety-related positive attitude and behaviours will benefit supervisors by helping in maintaining a good safety record. However, similar to POS, high quality LMX indicates overall positive interactions between employees and supervisors reflected by a variety of aspects, not specifically in term of safety issues. The direct link between LMX and employees' safety performance

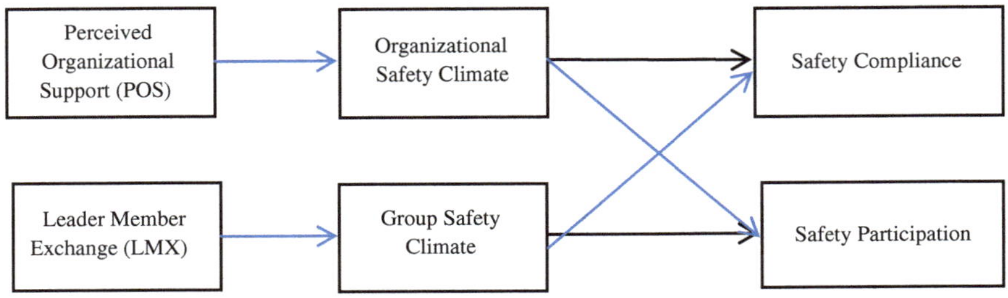

Fig. 8.1 POS, LMX, safety compliance and participation (authors' Figure)

may not be significantly strong and needs to be mediated by safety-specific constructs. This study proposes that the effect of LMX on employees' safety performance is mediated by group level safety climate. Zohar (2002) suggests a similar leadership-climate-injury mediation model, and argue that leader-member interactions influence supervisor's level of concern for subordinates' safety, which in turn affect employees' perception of safety climate in a workgroup. In a high quality leader-member exchange relationship characterized by norm of reciprocity, each party (i.e. supervisors and employees) facilitates the achievement of motivational goals of the other party, including personal welfare (Zohar 2002). Thus, high quality LMX is conductive to mutual concerns for each other. In a working environment inherent with high risk, high quality LMX will lead supervisors to concern for workers' safety, and put higher priority on safety than competing objectives such as production and cost. Supervisor's safety concern then contributes to workers' perception of positive safety climate in the workgroup, which in turn motivates employees to have improved safety performance. Based on the above argumentation, it is hypothesized as in Fig. 8.1.

8.5 Discussion and Conclusion

This paper proposes an integrative research model of predictors of construction workers' safety performance by integrating safety climate and social exchange perspectives. In the guidance of social exchange, two types of relationship-based leaderships are examined, i.e. perceived organizational support (POS) which represents employees' perception of the extent to which their organizations value them and care about their welfare, and leader member exchange (LMX) which refers to the quality of relationship between an employee and his/her supervisor. Previous studies have demonstrated that high levels of POS and LMX lead to employees' reciprocations of positive attitude and behaviours that benefit organizations and supervisors. In a working environment where safety is highly concerned (e.g. construction site), the reciprocations may also include safety-related positive attitude and behaviours. However, because POS and LMX reflect organizations and supervisors' general care and support to employees in a variety of aspects, the direct link between POS/LMX and workers' safety performance may not be significantly strong. This study suggests that the influence of POS and LMX on workers' safety performance is meditated by safety climate. Specifically, POS influence workers' safety performance through the mediating effect of organizational safety climate, which is mainly related to organizational safety policies and procedures; LMX influences workers' safety performance through the mediating effect of group level safety climate, which reflects supervisory safety leadership. Due to different natures of the two exchange relationships, this study also hypothesizes that POS is more associated with employees' safety compliance while LMX contributes more to employees' safety participation.

This study responds to recent researchers' argument that safety climate studies should be extended to explore relationships between safety climate and other influential constructs. It advances safety climate research by examining perceived organizational support (POS) and leader member exchange (LMX) as antecedents to safety climate, and explore how these relationship-based leaderships affect workers' safety performance through the mediating effect of safety climate. Although previous studies have attempted to examine the mediating role of safety climate, they only study safety climate at a single level, i.e. organizational safety climate (DeJoy et al. 2004, 2010) or group safety climate (Wallace et al. 2006). This study addresses multi-level safety climates identified in the construction industry, and attempts to explore the dynamic relationship between relationship-based leaderships, safety climate, and construction workers' safety performance at both organizational and group levels. It is expected that the conceptual research model proposed in this study provides implications for construction safety management researchers to further explore the interactive effect of leaderships and safety climate on workers' safety behaviours.

References

Atkinson, A., & Westall, R. (2010). The relationship between integrated design and construction and safety on construction projects. *Construction Management & Economics, 28*(9), 1007–1017.

Behm, M., & Schneller, A. (2012). Application of the Loughborough construction accident causation model: A framework for organizational learning. *Construction Management and Economics*, 1–16. doi:10.1080/01446193.2012.690884.

Blau, P.M. (1964) *Exchange and Power in Social Life*. New York: Wiley.

Borman, W. C., & Motowidlo, S. J. (1993). Expanding the criterion domain to include elements of contextual performance. In N. Schmitt & W. C. Borman (Eds.), *Personnel selection in organizations*. San Francisco: Jossey-Bass.

Bowen, D. E., & Schneider, B. (1993). The service organization: Human resources management is crucial. *Organizational Dynamics, 21,* 39–52.

Brondino, M., Silva, S. A., & Pasini, M. (2012). Multilevel approach to organizational and group safety climate and safety performance: Co-workers as the missing link. *Safety Science, 50*(9), 1847–1856. doi:10.1016/j.ssci.2012.04.010.

Clarke, S. (2006a). The relationship between safety climate and safety performance: A meta-analytic review. *Journal of Occupational Health Psychology, 11*(4), 315–327.
Clarke, S. (2006b). Safety climate in an automobile manufacturing plant. *Personnel Review, 35*(4), 413.
Cook, K. S., Cheshire, C., Rice, E. R. W., & Nakagawa, S. (2013). *Handbook of social psychology*. New York: Springer.
Cooper, M. D., & Phillips, R. A. (2004). Exploratory analysis of the safety climate and safety behavior relationship. *Journal of Safety Research, 35*(5), 497–512. doi:http://dx.doi.org/10.1016/j.jsr.2004.08.004.
DeJoy, D. M., Della, L. J., Vandenberg, R. J., & Wilson, M. G. (2010). Making work safer: Testing a model of social exchange and safety management. *Journal of Safety Research, 41*(2), 163–171. doi:http://dx.doi.org/10.1016/j.jsr.2010.02.001.
DeJoy, D. M., Schaffer, B. S., Wilson, M. G., Vandenberg, R. J., & Butts, M. M. (2004). Creating safer workplaces: Assessing the determinants and role of safety climate. *Journal of Safety Research, 35*(1), 81–90. doi:http://dx.doi.org/10.1016/j.jsr.2003.09.018.
Dement, J. M., & Lipscomb, H. (1999). Workers' compensation experience of north carolina residential construction workers, 1986–1994. *Applied Occupational and Environmental Hygiene, 14*(2), 97–106.
Eisenberger, R., Huntington, R., Hutchison, S., & Sowa, D. (1986). Perceived organizational support. *Journal of Applied Psychology, 71*(3), 500–507.
Fabrega, V., & Starkey, S. (2001). Fatal occupational injuries among hispanic construction workers of texas, 1997 to 1999. *Human and Ecological Risk Assessment: An International Journal, 7*(7), 1869–1883.
Fang, D., Chen, Y., & Wong, L. (2006). Safety climate in construction industry: A case study in Hong Kong. *Journal of Construction Engineering and Management, 132*(6), 573–584.
Flin, R., Mearns, K., O'Connor, P., & Bryden, R. (2000). Measuring safety climate: Identifying the common features. *Safety Science, 34*(1–3), 177–192. doi:10.1016/s0925-7535(00)00012-6.
Gillen, M., Baltz, D., Gassel, M., Kirsch, L., & Vaccaro, D. (2002). Perceived safety climate, job demands, and coworker support among union and nonunion injured construction workers. *Journal of Safety Research, 33*(1), 33–51. doi:http://dx.doi.org/10.1016/S0022-4375(02)00002-6.
Gouldner, A. W. (1960). The norm of reciprocity: A preliminary statement. *American Sociological Review, 25*(2), 161–178. doi:10.2307/2092623.
Griffin, M. A., & Neal, A. (2000). Perceptions of safety at work: A framework for linking safety climate to safety performance, knowledge, and motivation. *Journal of Occupational Health Psychology, 5*(3), 347.
Haslam, R. A., Hide, S. A., Gibb, A. G. F., Gyi, D. E., Pavitt, T., Atkinson, S., et al. (2005). Contributing factors in construction accidents. *Applied Ergonomics, 36*(4), 401–415.
Hinze, J. W., & Teizer, J. (2011). Visibility-related fatalities related to construction equipment. *Safety Science, 49*(5), 709–718.
Hofmann, D. A., & Morgeson, F. P. (1999). Safety-related behavior as a social exchange: The role of perceived organizational support and leader–member exchange. *Journal of Applied Psychology, 84*(2), 286.
Hofmann, D. A., Morgeson, F. P., & Gerras, S. J. (2003). Climate as a moderator of the relationship between leader-member exchange and content specific citizenship: Safety climate as an exemplar. *Journal of Applied Psychology, 88*(1), 170.
Holt, A. S. (2001). *Principal of construction safety*. Oxford: Blackwell Science Limited.
Huang, Y.-H., Ho, M., Smith, G. S., & Chen, P. Y. (2006). Safety climate and self-reported injury: Assessing the mediating role of employee safety control. *Accident Analysis & Prevention, 38*(3), 425–433. doi:http://dx.doi.org/10.1016/j.aap.2005.07.002.
Irumba, R. (2014). Spatial analysis of construction accidents in Kampala. *Uganda Safety Science, 64*, 109–120.
Kath, L. M., Marks, K. M., & Ranney, J. (2010). Safety climate dimensions, leader-member exchange, and organizational support as predictors of upward safety communication in a

sample of rail industry workers. *Safety Science, 48*(5), 643–650. doi:http://dx.doi.org/10.1016/j.ssci.2010.01.016.

Li, R. Y. M. (2009). Achieving compliance with environmental health-related land use planning conditions in Hong Kong: Perspectives from traditional motivation theories. *Journal of Environmental Health, 72*(4), 22–25.

Li, R. Y. M. (2011). *Game theory analysis on underground space planning knowledge sharing via web 2.0*. Paper presented at the 5th international conference on new trends in information science and service science, Macao.

Li, R. Y. M., & Poon, S. W. (2009a). Future motivation in construction safety knowledge sharing by means of information technology in Hong Kong. *Journal of Applied Economic Sciences, 4* (3(9)), 457–472.

Li, R. Y. M., & Poon, S. W. (2009b). Workers' compensation for non-fatal construction accidents: Review of Hong Kong court cases. *Asian Social Science, 5*(11), 15–24.

Li, R. Y. M., & Poon, S. W. (2011). Using web 2.0 to share the knowledge of construction safety as a public good in nature among researchers: The fable of economic animals. *Economic Affairs, 31*(1), 73–79.

Li, R. Y. M., & Poon, S. W. (2013). *Construction safety*. Berlin: Springer.

Liden, R. C., & Graen, G. (1980). Generalizability of the vertical dyad linkage model of leadership. *The Academy of Management Journal, 23*(3), 451–465. doi:10.2307/255511.

Ling, F. Y. Y., Liu, M., & Woo, Y. C. (2009). Construction fatalities in Singapore. *International Journal of Project Management, 27*(7), 717–726.

Lingard, H., Cooke, T., & Blismas, N. (2009). Group-level safety climate in the Australian construction industry: Within-group homogeneity and between-group differences in road construction and maintenance. *Construction Management & Economics, 27*(4), 419–432. doi: http://dx.doi.org/10.1080/01446190902822971.

Lingard, H., Cooke, T., & Blismas, N. (2010a). Properties of group safety climate in construction: The development and evaluation of a typology. *Construction Management & Economics, 28* (10), 1099–1112.

Lingard, H. C., Cooke, T., & Blismas, N. (2010b). Safety climate in conditions of construction subcontracting: A multi-level analysis. *Construction Management and Economics, 28*(8), 813–825. doi:10.1080/01446190903480035.

Lingard, H., Cooke, T., & Blismas, N. (2011). Coworkers' response to occupational health and safety: An overlooked dimension of group-level safety climate in the construction industry. *Engineering, construction and architectural management, 18*(2), 159–175. doi:http://dx.doi.org/10.1108/09699981111111139.

Lingard, H., Cooke, T., & Blismas, N. (2012). Do perceptions of supervisors' safety responses mediate the relationship between perceptions of the organizational safety climate and incident rates in the construction supply chain? *Journal of Construction Engineering & Management, 138*(2), 234–241. doi:10.1061/(ASCE)CO.1943-7862.0000372.

Mansor, N., & Ali, S. H. S. (2010). A B2C business agenda: Analyzing customers' perceptions towards bumiputera retailers. *Asian Social Science, 6*(7), 132–141.

Marchand, A., Simard, M., Carpentier-Roy, M.-C., & Ouellet, F. (1998). From a unidimensional to a bidimensional concept and measurement of workers' safety behavior. *Scandinavian Journal of Work, Environment & Health, 24*(4), 293–299. doi:10.2307/40966778.

Mearns, K., Whitaker, S. M., & Flin, R. (2003). Safety climate, safety management practice and safety performance in offshore environments. *Safety Science, 41*(8), 641–680.

Mearns, K. J., & Reader, T. (2008). Organizational support and safety outcomes: An un-investigated relationship? *Safety Science, 46*(3), 388–397. doi:http://dx.doi.org/10.1016/j.ssci.2007.05.002.

Meliá, J. L., Mearns, K., Silva, S. A., & Lima, M. L. (2008). Safety climate responses and the perceived risk of accidents in the construction industry. *Safety Science, 46*(6), 949–958.

Michael, J. H., Evans, D. D., Jansen, K. J., & Haight, J. M. (2005). Management commitment to safety as organizational support: Relationships with non-safety outcomes in wood

manufacturing employees. *Journal of Safety Research, 36*(2), 171–179. doi:http://dx.doi.org/10.1016/j.jsr.2005.03.002.
Mohamed, S. (2002). Safety climate in construction site environments. *Journal of Construction Engineering and Management, 128*(5), 375–384.
Mustapha, N., Ahmad, A., Uli, J., & Idris, K. (2011). Work-family facilitation and family satisfaction as mediators in the relationship between job demands and intention to stay. *Asian Social Science, 7*(6), 142–153.
Neal, A., & Griffin, M. A. (2006). A study of the lagged relationships among safety climate, safety motivation, safety behavior, and accidents at the individual and group levels. *Journal of Applied Psychology, 91*(4), 946–953.
Neal, A., Griffin, M. A., & Hart, P. M. (2000). The impact of organizational climate on safety climate and individual behavior. *Safety Science, 34*(1), 99–109.
O'Dea, A., & Flin, R. (2001). Site managers and safety leadership in the offshore oil and gas industry. *Safety Science, 37*(1), 39–57. doi:http://dx.doi.org/10.1016/S0925-7535(00)00049-7.
Settoon, R. P., Bennett, N., & Liden, R. C. (1996). Social exchange in organizations: Perceived organizational support, leader–member exchange, and employee reciprocity. *Journal of Applied Psychology, 81*(3), 219.
Siu, O.-L., Phillips, D. R., & Leung, T.-W. (2004). Safety climate and safety performance among construction workers in Hong Kong: The role of psychological strains as mediators. *Accident Analysis and Prevention, 36*(3), 359–366.
Smith, G. S., Huang, Y.-H., Ho, M., & Chen, P. Y. (2006). The relationship between safety climate and injury rates across industries: The need to adjust for injury hazards. *Accident Analysis and Prevention, 38*(3), 556–562.
Suárez-Cebador, M., Rubio-Romero, J. C., & López-Arquillos, A. (2014). Severity of electrical accidents in the construction industry in Spain. *Journal of Safety Research, 48*, 63–70.
Tam, C. M., Tong, T. K. L., & Chan, K. K. (2006). Rough set theory for distilling construction safety measures. *Construction Management and Economics, 24*(11), 1199–1206.
Tsui, A. S., Pearce, J. L., Porter, L. W., & Tripoli, A. M. (1997). Alternative approaches to the employee-organization relationship: Does investment in employees pay off? *The Academy of Management Journal, 40*(5), 1089–1121. doi:10.2307/256928.
Tyrie, A., & Ferguson, S. (2013). Understanding value from arts sponsorship: A social exchange theory perspective. *Arts Marketing: An International Journal, 3*(2), 131–153.
Vinodkumar, M. N., & Bhasi, M. (2010). Safety management practices and safety behaviour: Assessing the mediating role of safety knowledge and motivation. *Accident Analysis and Prevention, 42*(6), 2082–2093. doi:10.1016/j.aap.2010.06.021.
Waehrer, G. M., Dong, X. S., Miller, T., Haile, E., & Men, Y. (2007). Costs of occupational injuries in construction in the united states. *Accident Analysis Prevention, 39*(6), 1258–1266.
Wallace, J. C., Popp, E., & Mondore, S. (2006). Safety climate as a mediator between foundation climates and occupational accidents: A group-level investigation. *Journal of Applied Psychology, 91*(3), 681–688.
Wayne, S. J., Shore, L. M., & Liden, R. C. (1997). Perceived organizational support and leader-member exchange: A social exchange perspective. *Academy of Management Journal, 40*, 82–111.
Williamson, A. M., Feyer, A.-M., Cairns, D., & Biancotti, D. (1997). The development of a measure of safety climate: The role of safety perceptions and attitudes. *Safety Science, 25*(1–3), 15–27.
Zhang, P., & Ng, F. F. (2012). Attitude toward knowledge sharing in construction teams. *Industrial Management & Data Systems, 112*(9), 1326–1347.
Zohar, D. (2000). A group-level model of safety climate: Testing the effect of group climate on microaccidents in manufacturing jobs. *Journal of Applied Psychology, 85*(4), 587–596.
Zohar, D. (2002). Modifying supervisory practices to improve subunit safety: A leadership-based intervention model. *Journal of Applied Psychology, 87*(1), 156–163. doi:10.1016/s0925-7535(00)00014-x, 10.1037/0021-9010.78.2.306.

Zohar, D., & Luria, G. (2005). A multilevel model of safety climate: Cross-level relationships between organization and group-level climates. *Journal of Applied Psychology, 90*(4), 616.
Zohar, D. (2008). Safety climate and beyond: A multi-level multi-climate framework. *Safety Science, 46*(3), 376–387. doi:10.1016/j.ssci.2007.03.006.
Zohar, D. (2010). Thirty years of safety climate research: Reflections and future directions. *Accident Analysis & Prevention, 42*(5), 1517–1522. doi:http://dx.doi.org/10.1016/j.aap.2009.12.019.

MIX
Papier aus verantwortungsvollen Quellen
Paper from responsible sources
FSC® C105338

If you have any concerns about our products,
you can contact us on
ProductSafety@springernature.com

In case Publisher is established outside the EU,
the EU authorized representative is:
Springer Nature Customer Service Center GmbH
Europaplatz 3, 69115 Heidelberg, Germany

Printed by Libri Plureos GmbH
in Hamburg, Germany